ANALYSIS OF SURFACE WATERS

ELLIS HORWOOD BOOKS IN
WATER AND WASTEWATER TECHNOLOGY

This collection of authoritative works reflects the awareness of the importance to the world of water, wastewater treatment, and closely-related subjects. The titles are written and edited by experts from a wide range of countries closely concerned with research and development, monitoring and improving water quality and supplies, and treatment and disposal of wastewater.

Titles published in collaboration with the
WATER RESEARCH CENTRE, UK

AQUALINE THESAURUS 2
Editor: G. BOTHAMLEY, Water Research Centre
SEWAGE SLUDGE STABILISATION AND DISINFECTION
Editor: A. M. BRUCE, Water Research Centre
STABILISATION, DISINFECTION AND ODOUR CONTROL IN SEWAGE SLUDGE TREATMENT
Editors: A. M. BRUCE and E. S. CONNOR, Water Research Centre
BULKING OF ACTIVATED SLUDGE
Editors: B. CHAMBERS and E. J. TOMLINSON, Water Research Centre
BIOLOGICAL FLUIDISED BED TREATMENT OF WATER AND WASTEWATER
Editors: P. F. COOPER, Water Research Centre, and B. ATKINSON, University of Manchester
Institute of Science and Technology
WATER RESEARCH TOPICS
Editor: I. M. LAMONT, Water Research Centre
ENVIRONMENTAL PROTECTION: Standards, Compliance and Costs
Editor: T. J. LACK, Water Research Centre
SMALL WATER POLLUTION CONTROL WORKS
E. H. NICOLL, Scottish Development Department, Edinburgh
EFFECTS OF LAND USE ON FRESH WATERS
Editor: J. SOLBÉ, Water Research Centre
RIVER POLLUTION CONTROL
Editor: M. J. STIFF, Water Research Centre

Other books in Water and Wastewater
ALKALINITY-pH CHANGES WITH TEMPERATURE FOR WATERS IN INDUSTRIAL SYSTEMS
A. G. D. EMERSON,Independent Consulting Technologist
CHEMICAL PROCESSES IN WASTE WATER TREATMENT
W. J. EILBECK, University College of North Wales, and
G. MATTOCK, Waste Water Treatment and Resource Recovery Consultant
ANALYSIS OF SURFACE WATERS
H. HELLMANN, Federal German Institute of Hydrology, Koblenz
WATER-ABSTRACTION, STORAGE, TREATMENT, DISTRIBUTION
J. JEFFERY, General Manager, North Surrey Water Company
POLLUTION CONTROL AND CONSERVATION
Editor: M. KOVACS, Hungarian Academy of Sciences
NATURAL WATER SAMPLING
J. KRAJCA, Research Insitute of Geological Engineering, Brno, Czechoslovakia
HANDBOOK OF WATER PURIFICATION Second Edition
W. LORCH, The Lorch Foundation, Buckwood
BIOLOGY OF SEWAGE TREATMENT AND WATER POLLUTION CONTROL
K. MUDRACK, University of Hanover, FRG, and S. KUNST, Institute for Sanitary Engineering, Hanover, FRG
GROUND WATER PROTECTION
V. PELIKAN, Research Institute of Geological Engineering, Brno, Czechoslovakia
IMPROVING WATER QUALITY
M. RAPINAT, Compagnie Générale des Eaux, France
DRINKING WATER MATERIALS: Field Observations and Methods of Investigation
D. SCHOENEN and H. F. SCHÖLER, Hygiene-Institut der Universität, Bonn, FRG
HANDBOOK OF LIMNOLOGY
J. SCHWOERBEL, University of Freiburg, West Germany
BIOTECHNOLOGY OF WASTE TREATMENT AND EXPLOITATION
Editors: J. M. SIDWICK and R. S. HOLDOM, Watson Hawksley, Buckinghamshire

ELLIS HORWOOD BOOKS IN
AQUACULTURE AND FISHERIES SUPPORT

Series Editor: Dr. L. M. LAIRD, University of Aberdeen
BACTERIAL FISH PATHOGENS: Disease in Farmed and Wild Fish
B. AUSTIN and D. AUSTIN, Heriot-Watt University
FISH FARMING: Salmon and Trout
L. M. LAIRD and T. NEEDHAM, Fish Technology and Information Services, Aberdeen
MARINE FISH FARMING: Warm Water Marine Culture
J. F. MUIR, University of Stirling

ANALYSIS OF SURFACE WATERS

HUBERT HELLMANN
Chief of the Water Pollution Division
Federal German Institute of Hydrology, Koblenz

Translator: B. D. HEMMINGS
Translation Editor: E. B. PIKE

ELLIS HORWOOD LIMITED
Publishers · Chichester

Halsted Press: a division of
JOHN WILEY & SONS
New York · Chichester · Brisbane · Toronto

First published in 1987 by
ELLIS HORWOOD LIMITED
Market Cross House, Cooper Street,
Chichester, West Sussex, PO19 1EB, England
The publisher's colophon is reproduced from James Gillison's drawing of the ancient Market Cross, Chichester.

Translated from original Germand Edition *Analyhk von Oberflächengewässern ITAL*
Published in 1987 by Georg Thieme Verlag, Postfach 732, D-7000 Stuttgart 1

Distributors:

Australia and New Zealand:
JACARANDA WILEY LIMITED
GPO Box 859, Brisbane, Queensland 4001, Australia

Canada:
JOHN WILEY & SONS CANADA LIMITED
22 Worcester Road, Rexdale, Ontario, Canada

Europe and Africa:
JOHN WILEY & SONS LIMITED
Baffins Lane, Chichester, West Sussex, England

North and South America and the rest of the world:
Halsted Press: a division of
JOHN WILEY & SONS
605 Third Avenue, New York, NY 10158, USA

British Library Cataloguing in Publication Data
Hellmann, Hubert
Analysis of surface waters. —
(Ellis Horwood series in water and wastewater technology)
1. Water — Analysis
I. Title
551.48   QD142
Library of Congress Card No. 87–21349
ISBN 0–7458–0213–3 (Ellis Horwood Limited)
ISBN 0–470–20924–0 (Halsted Press)
Phototypeset in Times by Ellis Horwood Limited
Printed in Great Britain

Table of contents

Foreword

Among the very many diverse fields of analysis the examination of environmental samples involves the greatest risk of considering only the data instead of the necessary facts. In this area of investigation one should never overlook the truth that processes occurring in the environment occupy a continuum in space and time.

Surface waters and, to a very special degree, flowing waters demonstrate beyond all doubt that sampling must be performed within a space/time context in order to derive meaningful facts. Many investigations — even recent ones — fail to take this prerequisite sufficiently into consideration.

The author's opening comments show quite clearly that what is to come is no superficial, ill-considered or irrelevant treatment of the subject. The most important provisos for a proper and meticulous presentation of results for surface water investigations are observed, and their significance for the interpretation of data discussed. The factors to be considered are not merely confined to the water body itself but also take in its interaction with the atmosphere as well as with soils and sediments. Classification of the subject matter is reflected by the treatment of specific materials-related questions in separate chapters.

In reading the book it is abundantly clear that the author's treatment is based on many years of practical experience and investigation of the subject. As a result of the very concentrated involvement of the author in the study of the Rhine, many of the examples and conclusions are drawn from this waterway. The presentation is such, however, that in many cases, and above all in respect of the most important principles, the findings are applicable to other surface water systems.

In view of the ever more intensive efforts towards the protection of the quality of natural waters, relevant investigational concepts and analytical findings are steadily gaining in importance. This volume is thus very warmly recommended to those persons who are directly concerned with water quality monitoring and water quality protection and are thereby involved in the evaluation of trends in the quality of surface waters.

Leverkusen
Summer 1986   Herwig Hulpke

Preface

Without doubt the task of an analyst can be said to consist primarily of examining a given matrix, namely the sample in question, with the aid of the appropriate analytical methods. There is also general agreement that statistical methods are desirable for the purpose of validating analytical results. In the field of scientific analysis of natural waters, however, and in other fields as well, the task in hand cannot be said to be finished when the results exist merely in the form of a set of analytical data. For then — and not just since the emergence of socially and politically motivated environmental consciousness in the sixties — the question arises: 'What is the actual meaning of the results?' Hence there is an implied necessity for the analyst or the scientist responsible for evaluating the data to scan the fields peripheral to the analysis, to establish interrelationships and to a large extent think along interdisciplinary channels. In this connection certain basic and important factors are implicit in the task of water analysis, which must receive due attention if meaningful conclusions are to be reached.

Now that the numerous departments of what was formerly regarded as the science of natural waters are beset with some very pressing and down-to-earth issues of great political significance, one no longer speaks merely of the science of natural waters, but of water quality protection, water resource management, or even water transport management. These heavily publicised aspects of water analysis, however, benefit, as always, from the application of objective and unemotional scientific thinking, providing a framework into which the results of 'pure' analysis can be fed and incorporated within the context of the hydrological situation, and assessed with reference to the corresponding input from hydrobiological and hydrogeological disciplines.

The practical interpretation of the data should, however, be undertaken by the analyst himself, who is in by far the best position to gauge both the origin of the sample and the reliability of the data, among other things.

Viewed in this way the analysis of surface waters can be seen to be a very responsible task. The wide span of activity encompassed between the act of sampling and the final evaluation of the results will be illustrated with the aid of numerous examples, chiefly drawn from the author's own day-to-day experience. The fact that the information is drawn largely, if not exclusively, from one's own institute is not to be ignored — and here the beneficial collaboration of all the author's colleagues is gratefully acknowledged — but this does not imply that valuable research by other colleagues at home and abroad has been disregarded. Thus it will be seen that the purpose of this book is not to provide a treatise on the composition of waters in general or to assess their quality, nor even to describe the analytical techniques, but rather to highlight certain problems arising from an analysis and interpretation of data pertaining to surface water systems.

Nevertheless the book does give a passing survey of the level of pollution to which certain rivers, especially the Rhine, have been subjected during the last two decades. Although the layout has been determined largely by pedantic considerations, it also provides a snapshot of current events.

EDITORIAL NOTE

To assist English readers, the author has provided, on the following two pages, a map of the Rhine basin showing other major West German rivers, and diagrams of the location of sampling stations referred to in the text.

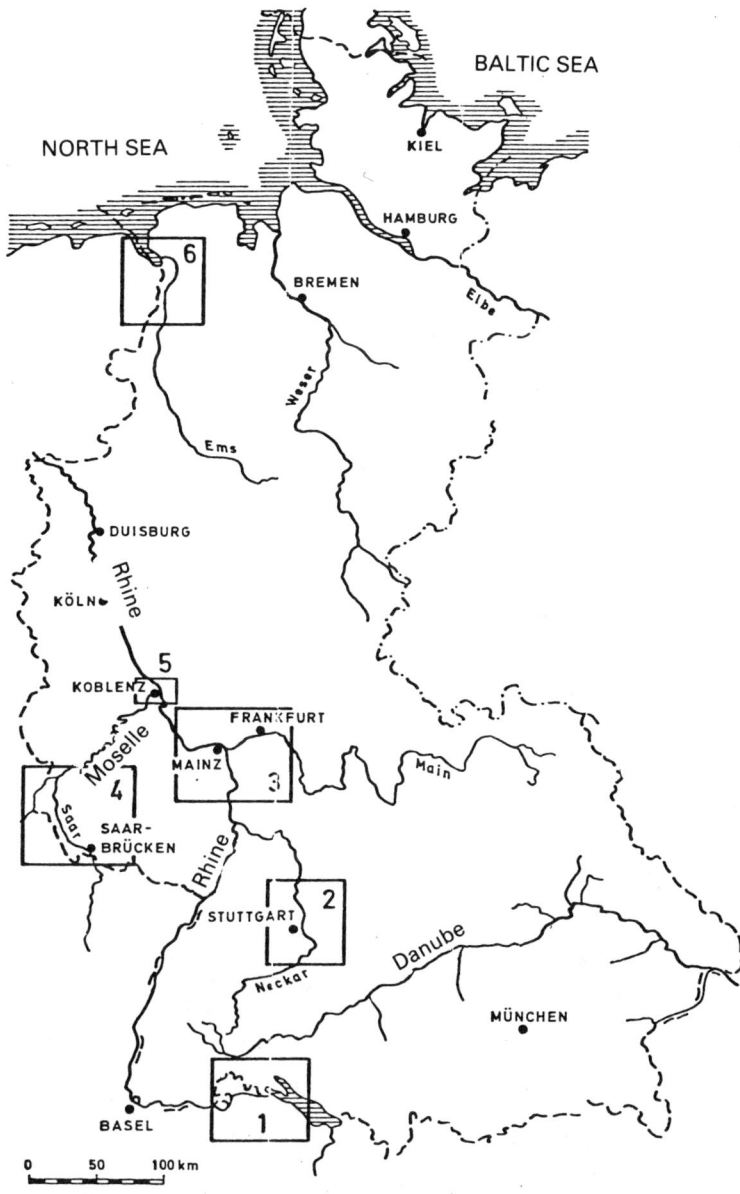

Fig 1a.

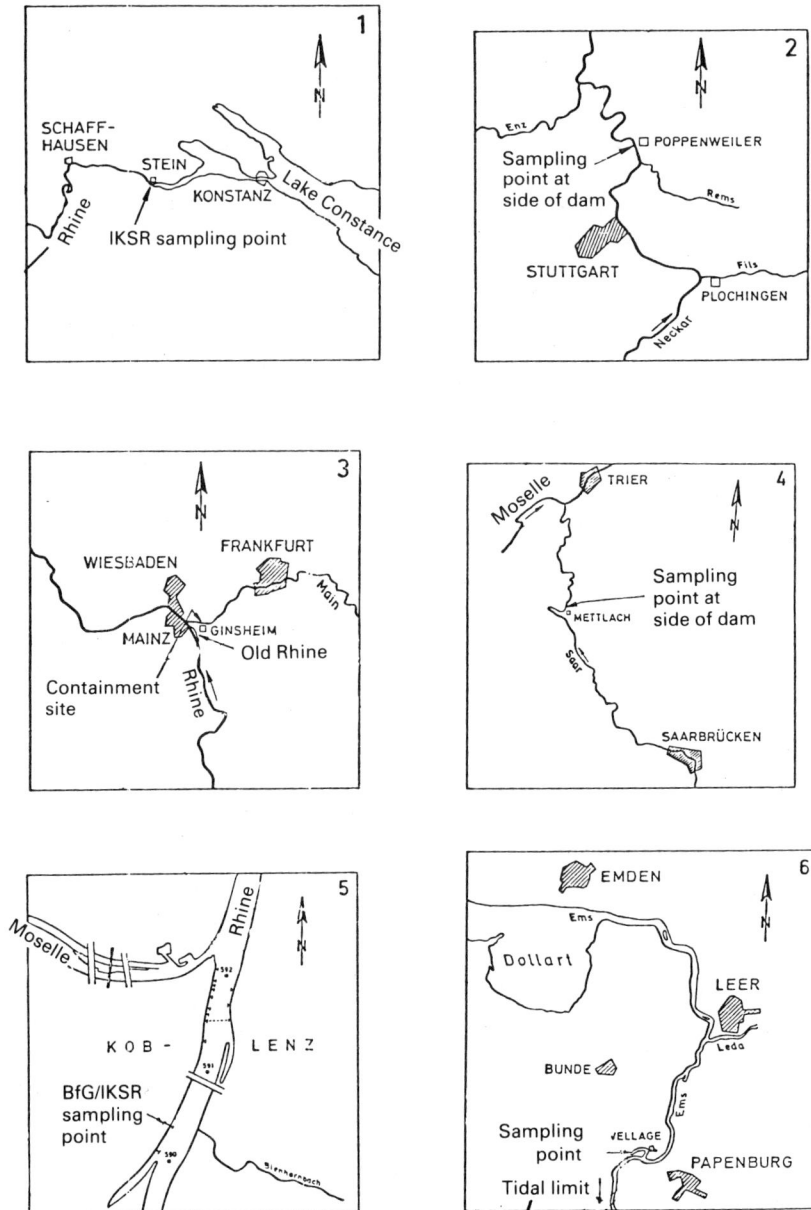

Fig. 1b.

Part I

General Principles

1

Hydrological principles of surface water analysis

1.1 THE DISCHARGE

Without question, it is possible to carry out the analysis of surface waters within a relatively narrow context, which only requires the concentration of the determinand to be ascertained. This concentration may even have predictive value under certain conditions, which only become apparent outside the scope of the analytical process. Examples of such a narrowly defined significance are water quality standards, the classification of natural waters according to use, or the emission standards that have assumed such political significance of recent years. The analytical result is assessed with reference to a predetermined 'scale' and has thus fulfilled its purpose.

The question of how such standards and limiting values are arrived at may, for the time being, be left on one side.

Only when the mass flow of a particular constituent must be calculated, and more than ever where trends are to be estimated, does the discharge, variously described as the water flowrate, throughput, etc., become a quantity of 'decisive importance', quite apart from other properties which distinguish it. It is certainly illuminating to realise that any quantity of effluent will be diluted to a greater or lesser extent in a water body and thus that every accountable substance will undergo continuous fluctuations in its concentration even though the input remains constant in all respects.

The fact that rivers seem to become cleaner in times of high rates of flow depends on just this dilution process. Whether a more acceptable water quality rating is thereby achieved remains a matter of some doubt.

If one considers, for example, the mean daily flows of the Rhine at Kaub in April 1983 (which correspond very closely with those at Koblenz) then the discharge increased from *ca.* 2000 m^3/s at the beginning of the month to a value greater than 6000 m^3/s, followed by a fall to around 2500 m^3/s. The range of dilution ratios of 1:3 for a hypothetical constituent within this period is far exceeded in the Moselle, where the spread reaches the level of 1:6.

Analytical results which are obtained in the presence of discharge fluctuations of this magnitude must therefore be very carefully considered before arriving at any interpretation. The classical example of the 'month kind to analysts' was October 1983 when the water flowrate remained between 1000 and 1200 m³/s for the whole month, with only slight variation, and was thus ideally suited to studies of water quality in a longitudinal profile.

Besides some years with relatively smooth discharge characteristics there are others which exhibit pronounced fluctuations, such as 1982. In retrospect it was exceedingly difficult to select representative data or even to carry out a trend analysis (Fig. 1.4).

On the basis of many years' experience it seems that the importance of discharge in the analysis of surface waters may indeed be verbally acknowledged yet only very few colleagues are acquainted with the relevant discharge values, their ranking and the connection between the rates of flow and the analytical results.

According to Figs. 1.1 to 1.3 the middle reaches of the Rhine exhibited discharges which ranged from 1000 to 6000 m³/s and the Moselle values from 500 to 3000 m³/s. Further aids to the ranking of these values are given by the standard averages of general use for flowing waters, calculated on a yearly basis.

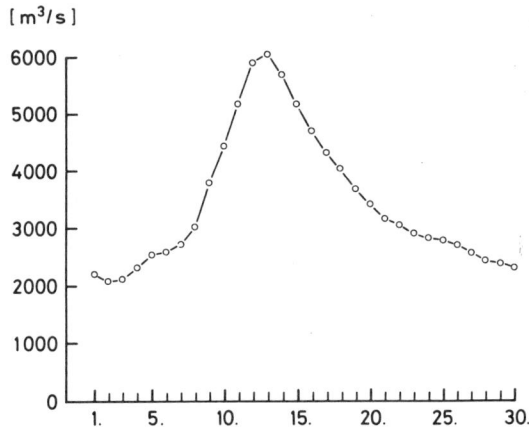

Fig. 1.1 — Discharge of the Rhine at Kaub: daily averages for April 1983.

The *average rate of flow* (MQ) is the average of all daily discharge values, the so-called daily means. It can be calculated over a period of a month, a year or several years. The yearly average value for the entire period from 1936 to 1965 at the Kaub gauging station was 1560 m³/s (Table 1.1).

The *average minimum flow* (MNQ) is obtained by averaging the lowest (e.g. for a period of a year) discharge values, and the *average maximum flow* (MNQ) is by analogy the arithmetic mean of the highest discharge readings from a succession of yearly records. The corresponding lowest (NNQ) or highest (HHQ) values must be accompanied by the appropriate year series. The lowest discharge in a certain period

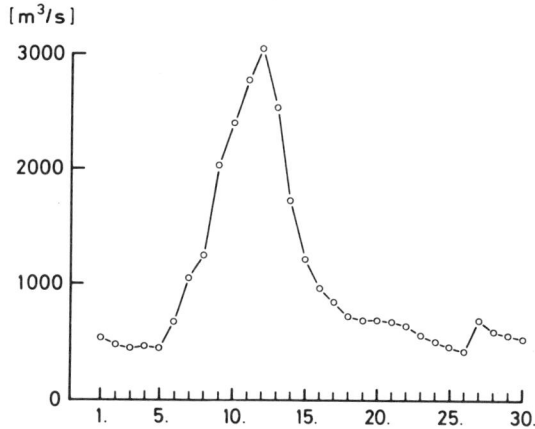

Fig. 1.2 — Discharge of the Moselle at Trier: daily averages for April 1983.

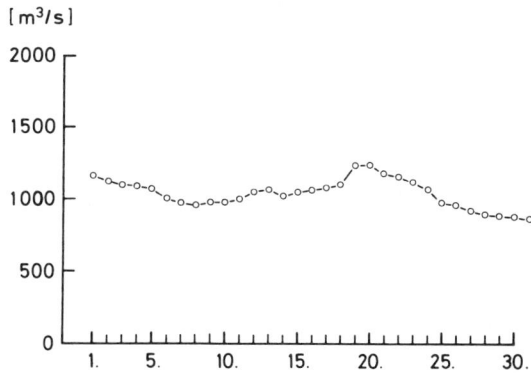

Fig. 1.3 — Discharge of the Rhine at Kaub: daily averages for October 1983.

of time is NQ; the highest recorded value is HQ. These 'principal values' provide a useful framework within which the actual values of discharge at any given time can be arranged. Thus an analysis of Rhine water obtained at Koblenz when the discharge was 950 m³/s must be regarded as pertaining to low flow conditions.

The magnitude of the discharge alone is not always sufficient for an expert interpretation of the analysis. A large catchment area is fed from many smaller catchments with differing levels of pollution. Thus for the Rhine in the broad sense there are two components to the discharge, that derived from the mountain districts and that from the uplands, inclusive of the Vosges district.

For the year 1967, which was in many respects representative of other years, the water level in the Rhine exhibited the following pattern in longitudinal profile.

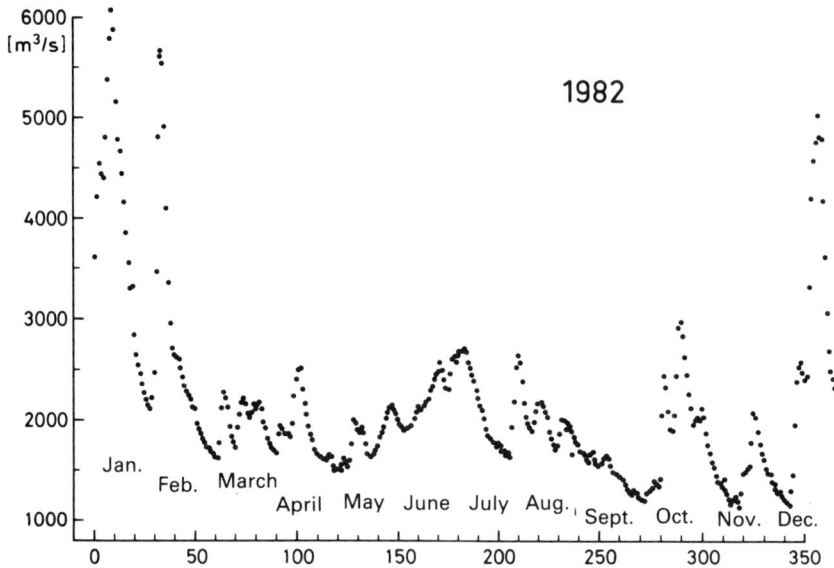

Fig. 1.4 — Discharge of the Rhine at Kaub during 1982, daily averages.

Table 1.1 — Discharge values for the Rhine at the Kaub gauging station and for the Moselle at Cochem (calendar years 1936–65). Discharge in m^3/s.

	Kaub[a]	Cochem[b]
MQ	1560	228
MNQ	702	58.9
MHQ	3950	1900
NQ	482	20
HQ	6150	3740

[a]546.2 km below the bridge across the Rhine at Constance, total catchment area 103 729 km^2.
[b]51.6 km above the confluence with the Rhine, catchment area 27 100 km^2.
Data from the German Water Data Book 1968.

In the winter quarter from December to March there was a large contribution from the uplands; the tendency to flood flows was strong. In the summer, melting of the Alpine snows resulted in a higher input from the Upper Rhine and the Aare, with a flatter peak between May and July. The peak was actually 'damped' by the Swiss lakes, including Lake Constance.

Pronounced low-flow periods for the Rhine occurred between October and the end of November. The inputs from both the mountainous and the upland catchments diminished considerably (Fig. 1.5).

Discharge [m³/s]

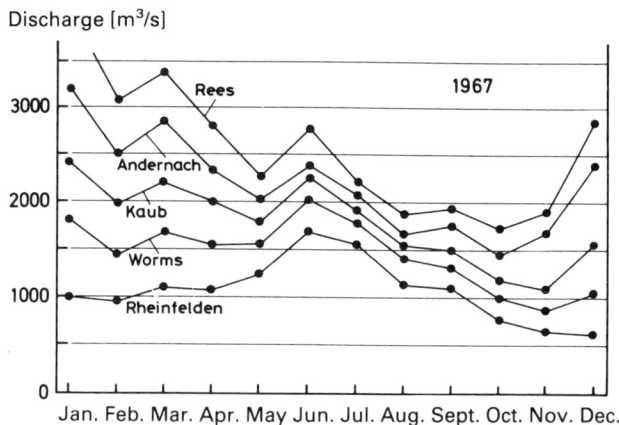

Fig. 1.5 — Discharge of the Rhine during 1967 at various gauging points.

1.2 DISCHARGE AND RAINFALL

Anyone who has experienced the commencement and subsequent progress of a flood event either in person or on the television screen will assert that any dilution of the pollutant load as a consequence of the increased discharge cannot occur — quite the contrary! The water looks dirtier. Of course the visual impression of a naturally turbid water (Chapter 2) should not be confused with anthropogenic or industrial contamination. This is something we shall revert to later.

Rapidly rising watercourses draw their flow principally from the runoff which occurs due to rainfall on the surrounding soil.

Rainfall records accompanied by a discharge profile which shows only slight relative displacement in time are illustrated in Fig. 1.6 and indicate a predominance of surface runoff. Yet the whole of the total rainfall will not follow this path.

The discharge profile of a surface water system is composed in general of three parts: the surface runoff, the interflow and the groundwater component (Fig. 1.7). In the case of heavy rainfall, the surface runoff increases very steeply at first; together with the impact of the raindrops on the soil surface, it gives rise to an appreciable soil erosion and transport effect. Part of the water, however, does not travel along the surface, but penetrates into the soil. Yet another fraction, termed the interflow, does eventually become runoff but only after a certain delay, and therefore helps to maintain the discharge at a higher level for some time. The more deeply infiltrating groundwater ultimately comes to the fore and constitutes the final phase of the discharge before this once again returns to its original value. Owing to the interaction of these three components there is a delayed decrease in the discharge in the wake of storm events.

In certain respects the process may be compared with the reading given by a galvanometer. The sudden increase would be followed by an equally sudden drop (aperiodic limiting case) were it not for the fact that a delayed exponential decrease occurs as a result of induction. This represents the leakage effect.

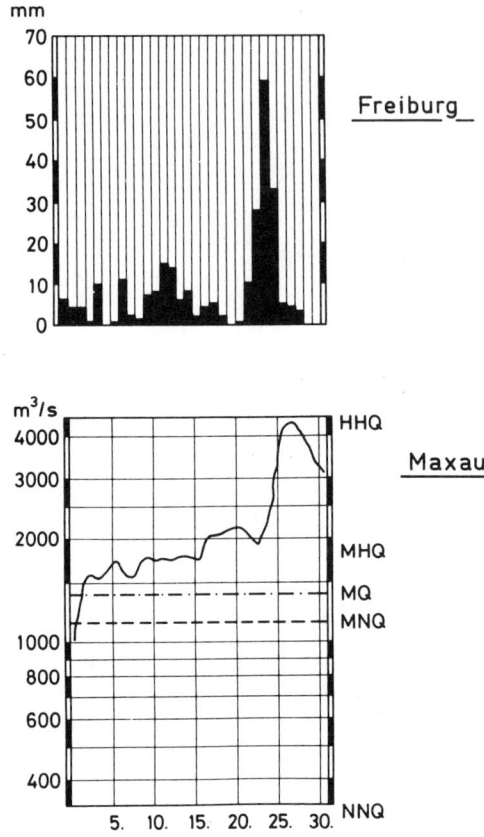

Fig. 1.6 — Relationship between rainfall and discharge showing rainfall at Freiburg
and discharge at Maxau (Karlsruhe) for May 1983.

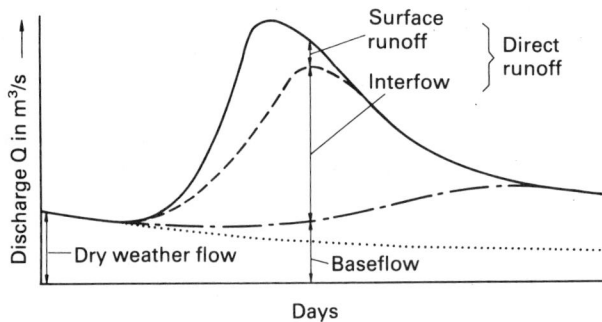

Fig. 1.7 — Composition of the discharge in response to rainfall — from DIN 4049, Part 1, Sect. 3.2.13.

The interflow cannot give rise to any surface entrainment, nor can the ground-water component.

The hydrologically distinguishable components can also be differentiated with respect to their widely differing content of chemical constituents. To this extent the interpretation of analytical data pertaining to flood flows is beset with special difficulties. Accordingly such an interpretation is possible only with certain reservations and limitations.

1.3 GROUNDWATER AND SURFACE WATER

Roughly four days after a rainstorm our German rivers revert almost entirely to the transport of groundwater, the so-called 'dry weather flow'. Apart from the spectacular flood events in certain years the groundwater flow contributes by far the largest proportion of the total flow. If one examines the published data, the equivalent runoff from the surface of the German Federal Republic, averaged over the course of a year, totals 313 mm. Of this figure the groundwater component contributes 254 mm, the surface runoff and the interflow 59 mm.

In chemical terms groundwater is composed of soil-filtered rainwater enriched with certain constituents during the passage through the soil and the deeper formations. Over and above this there exists in some places a close connection between groundwater and surface water.

In times of flood flows, the permeable, open-textured deposits in the vicinity of the river become filled with water, which is then released slowly at a later date (Fig. 1.8). Conversely groundwater derived from such flood plains contributes additional

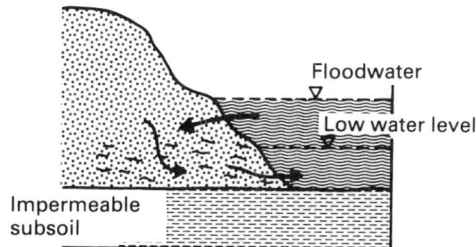

Fig. 1.8 — Interaction between groundwater and surface water flow.

components such as nitrate to rivers during periods of low flow, which will be detected during analysis of the surface flow. Depending on the agricultural or industrial purposes for which the surface water may be used, these are facts worthy of some reflection. Dangers to the groundwater and surface water systems may also arise as a consequence of the infiltration of leachate from landfill sites. Besides this, the particular situation created by the artificial raising of the water level (Moselle, Main) may be mentioned, in which case attention must be paid to the effect on boreholes in the vicinity of the river bank.

Thus the same river may, for the same flowrate, contain varying amounts of specific substances even at the same sampling point. The reason for this lies in the interactions between the river and the groundwater body in the region adjoining the channel.

1.4   THE CATCHMENT AREA

By the term catchment area one implies, according to DIN 4049 Part 1, No. 2.3, 'that area measured in a horizontal plane, from which water can flow to a particular point'. This area may be quite small, as for example for the Sieg, with F_N=748 km^2, relatively large (for the Moselle F_N=28 152 km^2) or very large indeed (for the middle Rhine F_N=103 729 km^2). The larger the catchment area the smoother as a rule is the pattern of discharge variation. Small catchment areas ordinarily signify small average discharge values. A specially favourable case is that of the Rhine, for which the lakes in the foothills of the Alps exert a buffering or damping effect on the discharge behaviour, as previously indicated.

Close to the Dutch–German border at Rees, the catchment area extends to a total of 159 683 km^2, which is more than half the area of the German Federal Republic. The discharge at this point can thus be equated on average and to a rough approximation with 50% of the input of surface water to the coastal inshore waters.

The analysis of surface waters is also bound up with the catchment area. A special situation arises if important portions of the catchment area lie outside the area open to immediate inspection. Since for the Moselle 66% of the catchment area lies in France and Luxembourg, and hence the predominant fraction of its anthropogenic pollution originates from those countries, the difficulty of correctly interpreting the results of analyses in isolation is all too obvious. The same applies to other rivers, such as the Upper Rhine, which have different countries adjoining the river on the two sides.

When one takes into account the geological characteristics of the catchment area, which are responsible for a certain baseline level of contamination, then one completes the catalogue of problems facing the analyst of surface waters from this point of view.

LITERATURE

DIN 4049 Teil 1; Hydrologie; Begriffe, quantitativ. Wechmann, A. (1964) Hydrologie, R. Oldenbourg-Verlag, München, Wien.

Rössert, R. (1969) Grundlagen der Wasserwirtschaft und Gewässerkunde, R. Oldenbourg-Verlag, München, Wien.

1.1

Liebmann, H. (1969) Der Wassergüteatlas, Methodik und Anwendung. Münchner Beiträge zur Abwasser-, Fischerei-und Flußbiologie, Bd. 15, R. Oldenbourg-Verlag, München.

Verordnung über Trinkwasser und Brauchwasser für Lebensmittelbetriebe (Trinkwasser-Verordnung) von 31. Januar 1975.

Grenzwerte für Oberflächenwasser. KfW-Mitteilungen 2, 1975.

Rheinwasserverschmutzung und Trinkwassergewinnung. Memorandum der Internationalen Arbeitsgruppe der Wasserwerke im Rheineinzugsgebiet (IAWR), Gas-Wasser-Abwasser 53 (1973) Nr. 6.

Dinkloh, L. (1982) Stand des Gewässerschutzes 1982 in der Bundesrepublik Deutschland aus der Sicht der Gestzgebung, Vom Wasser **59**, 17–37.

Deutsches Gewässerkundliches Jahrbuch; z. B. Rheingebiet, Abflußjahr 1968, Hrsg. Landesamt für Gewässerkunde Rheinland-Pfalz, Mainz.

Der Rhein. Ausbau, Verkehr, Verwaltung (1951) "Rhein" Verlagsgesellschaft mbH. Duisburg; dort: Eschweiler, W., Hydrographie des Rheinstroms, S. 34–61.

1.2

Mendel, G., Ubell, K. (1973) Der Abflußvorgang, Dtsch. Gewässerk. Mitt. **17**, H. 2, 33–39; H. 3, 85–91.

1.3

Hydrologischer Atlas der Bundesrepublik Deutschland, Textband, DFG (...) Kap. 5 Abfluß (H. J. Liebscher u. R. Keller); Kap. 10 Wasserbilanz der Bundesrepublik Deutschland (R. Keller).

Mendel, H. G. (1974) Korrelation von Niederschlag und Abfluß in drei Deutschen Repräsentativgebieten, Beitr. zur Hydrologie H. 2., 25–66.

Koppe, P. (1970) Grundlegende Überlegungen und Untersuchungen über die hydrochemischen Beziehungen zwischen Flußwasser und dem Wasser ufernaher Brunnen, Schriftenreihe des Vereins für Wasser-, Boden- und Lufthygiene **33**, 129–142.

Knöpp, H., Müller, D. (1980) Zur Beurteilung der Auswirkungen von Staumaßnahmen auf die Gewässergüte und die Qualität des Uferfiltrats, Z. Wasser Abwasser Forsch. **13**, 86–99.

Die chemischen Zusammenhänge zwischen Oberflächenwasser und Grundwasser im Moseltal zwischen Trier und Koblenz (1973). Besondere Mitteilungen zum Deutschen Gewässerkundlichen Jahrbuch Nr. 38, Hrsg. Bundesanstalt für Gewässerkunde, Koblenz.

Kußmaul, H. (1975) Untersuchungen über die hydrochemischen Zusammenhänge zwischen Flußwasser und Uferfiltrat am Niederrhein, in Uferfiltration — Bericht des BMI-Fachausschusses "Wasserversorgung und Uferfiltrat", Bundesministerium des Innern, Bonn, S. 151–168.

2

Solids contained in surface waters

2.1 SUSPENDED SOLIDS TRANSPORT

The discharge is a co-adjudicating factor in the interpretation of analytical results for surface water studies. It bears on the interpretation of values for concentration and mass flow of specific substances. In so far as one considered only those substances which are dissolved, e.g. chloride, nitrate, or possibly oxygen, the discharge is by far the most important external factor.

For other substances which are either partially or completely associated with the suspended solids, the so-called suspended solids transport (or particulate content) of the water assumes equal importance and constitutes a further decisive parameter for analytical purposes.

Suspended solids are defined in DIN 4049 Part 1 as 'solids which exist in equilibrium with the water or are maintained in suspension due to turbulence' (Para 3.5.13). Measurements show that the suspended solids content (S, in mg/l or g/m^3) increases or decreases with the discharge Q. The hypothesis that a correlation exists between these two quantities is confirmed by Fig. 2.1. Bogardi recognised such a

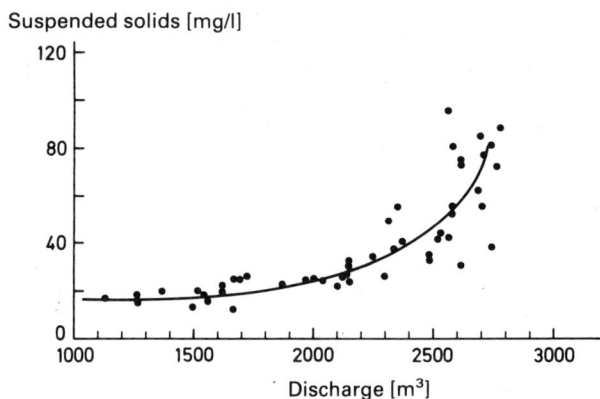

Fig. 2.1 — Correlation between suspended solids content and discharge for the Rhine at Koblenz, 1967.

relationship in the case of the Danube, of an exponential form, expressed to a first approximation by

$$S = aQ^b$$

He demonstrated that the hydraulic quantities alone did not govern the magnitude of the suspended solids transport, but that a greater influence was exerted by the quantity of suspended solids which could be supplied by the catchment area and the surface water system upstream from the measuring point. Bogardi successfully investigated the physical significance of the exponent b and its dependence on hydrogeological and geographical factors. The coefficient a could not be determined from physical criteria as it was independent of the discharge and the extent of the catchment area.

For calculating the values of the constants a and b for the Koblenz gauging station, the above equation was converted to logarithmic form:

$$\log S = \log a + b \log Q$$

After some corrections the results for the middle Rhine at Koblenz could be represented by

$$S = 4.1 \times 10^{-4} \times Q^{1.44}$$

This equation implies that the solids transport by the Rhine at Koblenz not only increases with the discharge, but does so at an even faster rate, namely as an exponential function of the discharge. For example, solids transport can be calculated for various levels of the discharge as follows:

NQ	482 m^3/s	10 g/m^3
MNQ	702 m^3/s	10 g/m^3
MQ	1560 m^3/s	20 g/m^3
MHQ	3950 m^3/s	71 g/m^3
HQ	6150 m^3/s	134 g/m^3

This may astonish those who regard the visible 'dirt' in flowing waters merely as a consequence of industrial development. Were this so, then an increase in discharge would bring about a dilution of the 'inevitable' contamination and the suspended solids content would become smaller, as is indeed the case for dissolved salts and organic substances in the water. The results, however, make it clear that by far the largest part of the ordinary suspended solids fraction (of inorganic composition) must have contaminated the Rhine from time immemorial. At the same time it should not be denied that the Rhine has become more polluted. The actual pollutants are, however, present mainly in the soluble form, for which the increased biogenic productivity of the water is evidence, as well as being combined with the suspended solids of natural origin.

The exponential increase of S with discharge according to the equation stated above is explicable not only in terms of the entrainment effect of runoff during rainfall. Besides the values obtained for the Rhine which agree most closely with the equation are those which pertain to periods with no rainfall, i.e. principally those relating to the snowmelt period in the Alpine districts in spring and early summer. The mathematical connection between suspended solids content and discharge for this region is thus principally dependent on the abrasion and erosion taking place on the river bed and the inputs from tributaries. The contribution from surface runoff is superimposed on these in a more or less regular fashion. The conditions under which the numerical value of b was determined are no longer fulfilled in the case of heavy and sudden rainfall events with their subsequent manifestations such as the carrying away of shoals, the denudation of fields and terraces, etc. One need not, however, belittle the importance of such a correlation, owing to the recognised fact that in such cases the solids transport invariably increases faster than the calculated profile and then diminishes again (Fig. 2.2). From a long-term viewpoint an exceptionally heavy

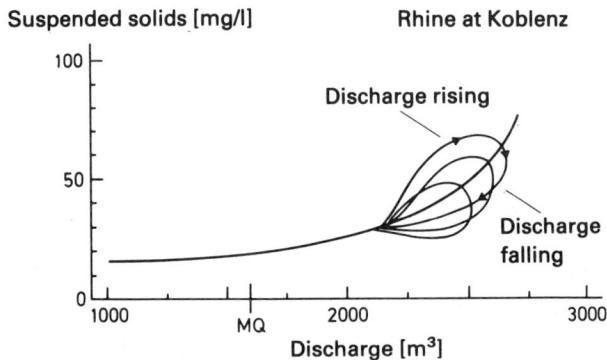

Fig. 2.2 — Suspended solids content vs discharge behaviour for a watercourse in response to rising and falling discharge (schematic).

wash-out with very high local concentrations of suspended solids becomes less significant. This applies even more as the size of the catchment, the average water flow and the length of the watercourse concerned increase. For a small tributary, however, the observation of Rinsum is correct that rainfall events which are the cause of the increased flow, of different type and duration and with very variable types of ground cover, do not give rise to suspended solids in amounts susceptible to representation by a mathematical formula. Understandably the same applies to the biological productivity of a river.

The equation derived for Koblenz was strictly speaking only valid for a single point on the river, likewise the equation $S = aQ^b$ in this general form must apply not only to a particular point in the water but must be valid for any point in the stream cross-section. It must be determined for each and every case, however, whether and to what extent the values of the exponent b and the coefficient a may vary.

Summarising, we can say that erosion as manifested by scour of bed material and the attrition of meanders and banks can be most readily correlated with discharge

(Fig. 2.1). Soil wash-out creates an additional load of suspended solids which can be represented graphically by the pear-shaped portion of Fig. 2.2. With increasing discharge very large suspended solids contents may be observed, while with a decreasing discharge very much lower values, even for the case of equal flows, will be recorded.

In an average discharge year, e.g. 1967, the superficial wash-out makes a substantially lower contribution, of about 0.2 to 0.4 million tonnes (Rhine at Koblenz), to the total suspended solids load than does erosion. In very wet years, however, the proportions may vary. From estimates based on our own incidental observations—the continuous records of the Federal Institute for the Science of Natural Waters are published in Year Books—the amount of soil flushed from the catchment area at Koblenz in 1966 amounted to about 3 million tonnes. Related to the total surface area of the catchment, at 110 074 km^2, this was equivalent to a decrease of 0.01 mm in soil cover. For comparison, the annual surface removal in the catchment of the Mississippi has been calculated as 0.033 mm; for this value it was, however, assumed that the entire suspended solids load represented surface soil removal.

Exceptional discharge conditions such as those occurring in the flood event lasting from 13 to 17 January 1955, may be the occasion of an extremely high level of wash-out. An examination of data by the Office of Waterways and Navigation at Stuttgart provided information on the amounts involved for the Neckar alone. This indicated that an amount of 3.2 million m^3 of solids entered the Rhine at Mannheim. According to another Office the January floods referred to were responsible for the transfer of 1.09 million tonnes of suspended solids from the Neckar into the Rhine. One can easily figure how huge the total load in the Rhine (which was finally transported into the North Sea) must have been during those four days.

These figures provide a good illustration of how the 'suspended solids situation' occupies a central place in the study of natural waters. By turning our attention to their chemical analysis, some further general details will become clear.

Chemical description

When in the course of the period 1965–1967 the chemical composition of the suspended solids load in the Rhine catchment became clear to the author, it occasioned a certain disillusionment, especially on account of the extensive similarity of the suspended solids composition. Broadly speaking it was possible to distinguish between an organic and an inorganic fraction.

The inorganic fraction consisted of the mineral components, chiefly crystalline in nature, but also contained the skeletons of diatoms. The organic fraction which was vaporised on ignition at 600°C consisted of plankton and detritus. The distinction between organic and inorganic fractions is governed by the analytical techniques and not merely by the organic and inorganic constituents actually present. This is particularly apparent from the fact that the skeletons of diatoms—although of organic origin—are included here among the 'inorganic' fraction, while on the other hand the water of crystallisation of clays is included with the loss on ignition and hence is reckoned with the organic fraction. The interpretation of results obtained in this way accordingly requires particular caution. The inaccuracies associated with the classification of the respective materials were, however, outweighed by the simplicity of the analytical procedures.

From chemical analysis of Rhine suspended solids, by far the largest fraction is inorganic in nature. For example, Table 2.1 shows the composition of the suspended

Table 2.1 — Composition of suspended solids

Constituent	SiO_2	Al_2O_3	Fe_2O_3	CaO	MgO	CO_2	K_2O	Na_2O
Proportion (%)	44.4	10.9	4.8	7.9	2.3	5.0	1.0	0.1
Total = 76.4%								

solids recovered from a 500 l sample of water taken from a point 80 cm below the water surface at Rhine km 590. The discharge was 1984 m^3/s.

Mineral description
According to expectations, the mineralogical examination of the river sediments provided evidence of the following minerals: calcite, pyrites and titanite. They were present in small quantities in the sediment from the Upper Rhine at Öhningen, less so in the Middle and Lower Rhine. For those regions higher amounts of quartzite were obtained.

The sediments examined contained a very high percentage of clay and silt; in the opinion of many authors the mineral resources of the Rhine are composed almost exclusively of illite, a substance similar to montmorillonite produced by degradation and weathering of the basement complex. Illite is formed primarily from micas, particularly from muscovite and biotite.

Results of chemical analyses of suspended solids at Koblenz were in good agreement with the mineralogical composition of the stated materials.

The universal distribution of the same inorganic material in different watercourses is supported by the results of further investigations. In the Lower Elbe, for example, besides the opaque minerals calcite, pyrites and titanite, the same components were found. In addition, montmorillonite and chloride minerals were present. For the Kiel canal the same was true. The identity of the inorganic fraction of the 'undissolved solids' in rivers with the inorganic constituents of our soils throws a revealing light on the origin of the material. The suspended solids cannot thus be equated with the nationwide problem of water pollution without further thought. The nearest thing to actual 'pollution' is the organic fraction.

Hydrobiological description
The suspended solids in flowing waters appear sometimes from a distance as a dark, more or less uniform turbidity of the water and hence give the impression of pollution. From a distance of a few metres, however, the apparently homogeneous 'pollution' resolves itself into a large number of small particles. The appearance of

the particles, which may be up to 2 cm in size, is further described by Schmidt-Ries. One distinguishes between plankton, detritus and tripton. According to DIN 4049 Part 2 Para 3.42, plankton is the term used for animal and vegetative organisms carried along by the water without more definite motility. Detritus, Para 3.46, is composed of the residues of organisms and colloidal matter originating, e.g., from domestic wastes. Tripton is the term applied to the suspended inorganic matter.

Lauterborn occupied himself from about the beginning of the century with the study of plankton in the Rhine. He stressed the importance of diatoms for the biological self-purification of the river and demonstrated that the tripton load can interfere with the environment of the plankton by absorption of heat and sunlight. Even at that time (1893–1905) discharges of sewage were associated with harmful effects on the river biocoenosis.

Marsson first made the observation (1906) that the amount of tripton substantially exceeded that of the plankton. According to the same author green algae were rarely present in the Rhine, but diatoms were correspondingly more abundant. Also a typical type of tripton in the Rhine was said to consist of a fine, soil-rich carbonate-based mineral mixture; the more loamy and clayey fractions were derived from the tributaries.

Conclusions

The suspended solids fractions must be taken into account in the analysis of surface waters by reason of their amount alone.

For longitudinal profile investigations, a true assessment of the analytical results is usually only possible if the transport of water and particulates is not upset by serious local disturbances, such as, for example, localised heavy rainfall. A typical result of measurements along the Rhine in 1967 is presented in Fig. 2.3. From this it can be seen that for an average discharge, a load of around 6000 t/d was transported across the border into Holland, of which about 30% was composed of organic matter. Examples of mean annual transport rates of suspended solids in German rivers are given in Table 2.2.

Besides the *quantity* of suspended solids, something should be said about their absorption capacity for numerous organic and inorganic trace substances. According to pertinent investigations, apart from the organic matter, the clay fraction is the principal adsorbant. However, the adsorptive properties of the average silt are not inconsiderable. The particle sizes of the suspended solids, except during periods of high flowrate, are generally situated within the clay and silt region (Fig. 2.4).

At one and the same point the general picture of suspended solids changes with the discharge (Fig. 2.5). While in the Middle and Lower Rhine as well as in similarly affected watercourses, under conditions of dry weather flow the form and size of the particles is governed by the plankton and the detritus, during and after the more severe rainfall events clay particles associated with wash-out make their appearance (Phase II, Fig. 2.5). Diminishing surface erosion under a sustained high level of discharge leads to dilution of the still relatively small particles (III), after which the nature of the particulate fraction then slowly returns to its original form and dimensions due to the growth of bacteria and plankton biomass.

The progress of events in a river is dynamic, not only in connection with the visible appearance of the suspended solids. At the surface of these solid particles and

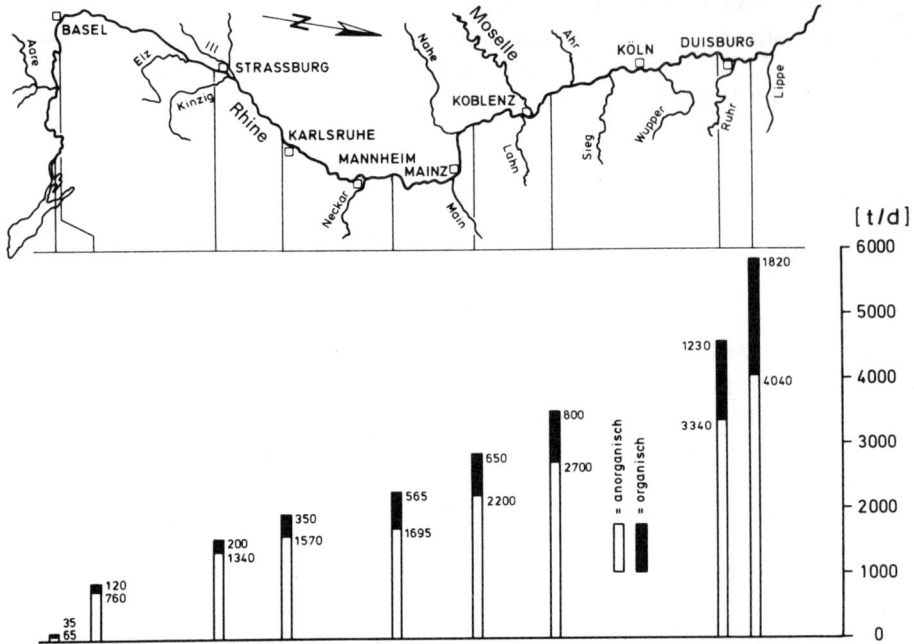

Fig. 2.3 — Suspended solids transport in the Rhine in 1967—a longitudinal profile from
Constance to Emmerich/Lobith under conditions of dry weather flow.

Table 2.2 — Mean annual mass transport rates for suspended solids at selected
locations

River and gauging point	Mean annual transport (× 1000 t)	Calendar period
Rhine, Rees	3400	1967/76
Neckar, Rockenai	240	1972/76
Main, Griesheim	110	1973/76
Moselle, Lehmen	450	1964/76
Weser, Intschede	370	1970/76
Elbe, Hitzacker	800	1964/76
Danube, Velshofen	510	1967/75

Source: Annual Report of the Federal Institute for Natural Waters (Bundesanstalt für Gewässerkunde),
1977.
   On average the coastal waters of the German Federal Republic receive a total of about 5 million tonnes
of suspended solids annually.

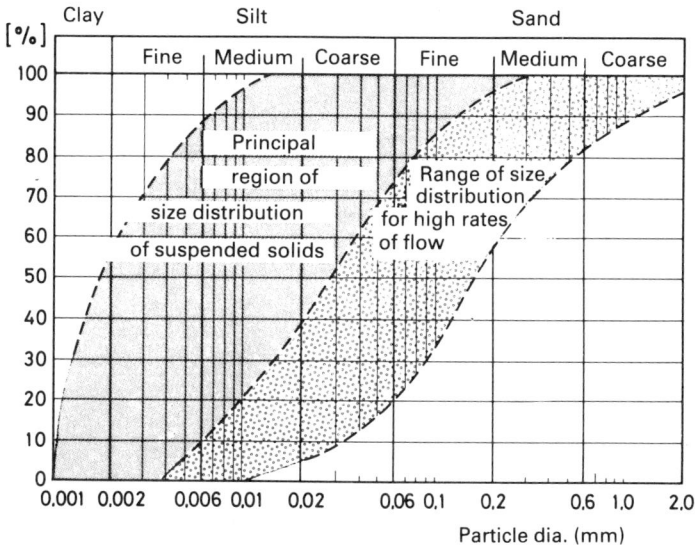

Fig. 2.4 — Particle size distribution of Rhine suspended solids.

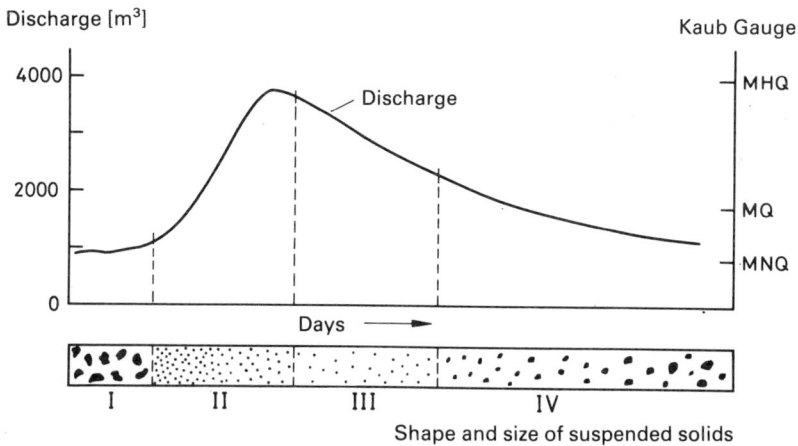

Fig. 2.5 — General picture of suspended particles typical of a river receiving sewage effluents under different flow conditions. I Dry weather flow; II Rising discharge period; III Falling discharge period; IV Dry weather flow.

the still smaller colloidal solids occur processes involving the adsorption and desorption of organic and inorganic trace substances which are invisible to our eyes and also little understood qualitatively and quantitatively.

2.2 EROSION, DEPOSITION AND SEDIMENT BUILD-UP

Suspended solids cannot be deposited at points where the flowing water causes erosion. One therefore does not find any sediment or, as frequently stated, sludge deposits in free-flowing waters. Deposits of suspended solids may, however, occur in very slow-flowing waters with regulated flow, e.g. in the Moselle, the Main and the Neckar as well as in the Upper Rhine (Iffezheim barrage), in retention ponds or other quiescent zones such as harbours, and during the retention of floodwaters by storage in flood meadows. On the whole the annual rate of sedimentation in the watercourses of the Federal Republic is relatively small. It ranges from 2% to a maximum of 10% of the total quantity of solids transported. The importance of these sediments in inland waters lies in their contamination by trace substances, about which more detail will be given later. They also influence the water quality and interfere with navigation either in the river channel or in harbours. In a regulated river, very low rates of flow may occur in the vicinity of the dam. For velocities of 0.20 m/s and below the suspended solids increasingly fall to the bottom and remain there until the next flood event (Fig. 2.6). As long as the deposits do not become

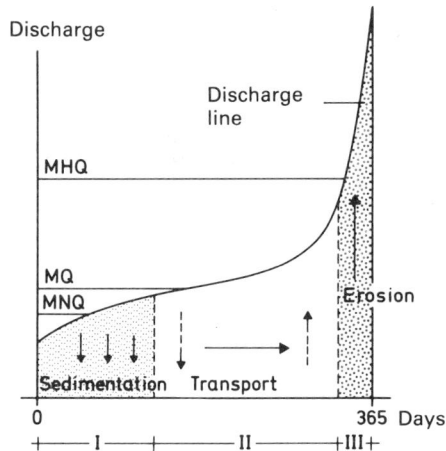

Fig. 2.6 — Behaviour of suspended solids in an impounded reach (schematic). Discharge line = number of days for which the actual discharge is below the level indicated.

consolidated, they can release individual particles, and again take the form of suspended solids, which helps to accentuate the exponential increase in the bedload already referred to, under conditions of high flowrates. Strongly compacted deposits must, however, be artificially removed from time to time. In many places they give rise to a disposal problem.

A classic example is provided by the very large-scale Poppenweil dam below Stuttgart and the confluence with the Rems. At this point from 30 000 to over 50 000 m^3 of solids are deposited annually, depending on the discharge situation, of which the largest portion is no longer capable of being re-entrained during flood flows. The water content of the sludge lies between 60% and 80% by weight. The largest quantity separates out beside the mole which divides the lock passage from the weir and the power station channel (Fig. 2.7). The principal flow in times of dry weather takes place in the direction of the power station (Fig. 2.8), so that no clayey

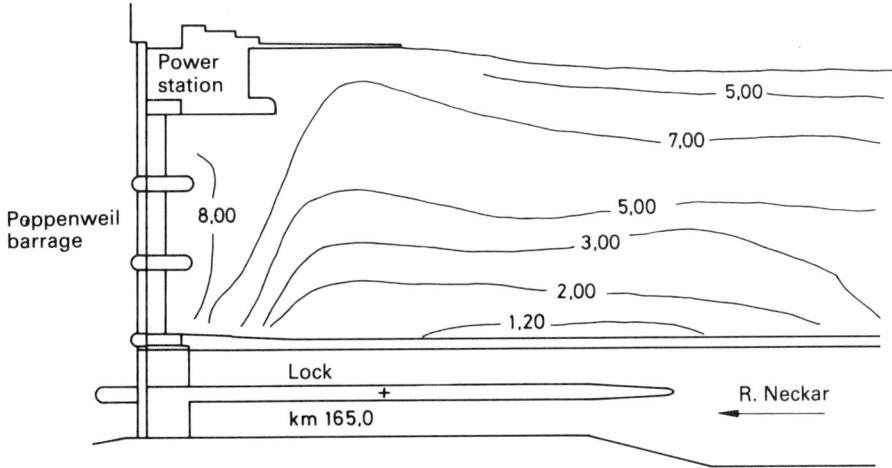

Fig. 2.7 — Distribution of sediment deposits in the Neckar directly behind the Poppenweil barrage, 1972 (figures indicate the water depth).

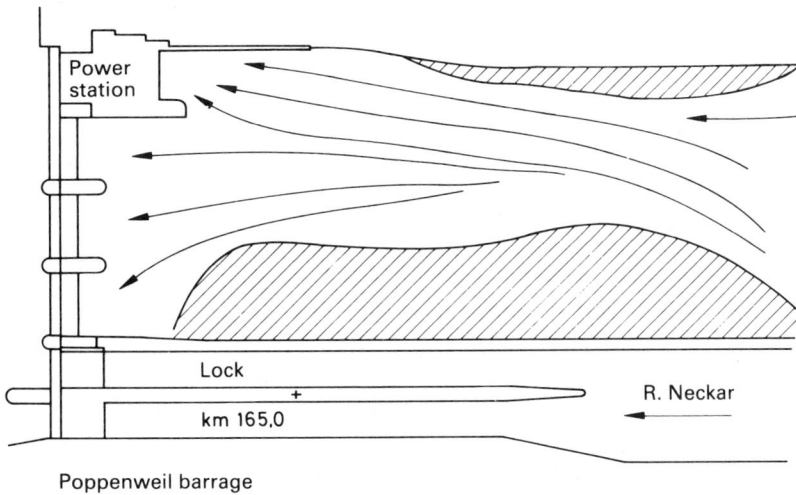

Fig. 2.8 — Flow patterns and sediment deposition behind the Poppenweil barrage on the Neckar (schematic).

or silty material can separate out along this path. Due to the decreasing flowrate, the suspended solids are fractionated according to their particle size. Sampling and interpretation of analytical results are particularly problematic in the case of sludge deposits of this kind (Chapters 5 to 7).

In inland lakes practically the entire load of suspended solids is deposited. There, too, the suspended solids and bedload are fractionated according to particle size. The clay fraction remains longest in suspension and is therefore carried farthest into the interior of the lake (see Fig. 7.14). In eutrophic lakes there is a layer of solids resulting from the more or less pronounced phytoplankton productivity. Frequently the resulting biomass becomes incapable of further decomposition, so that the sediments or sludge deposits on the floor of the lake are added to from year to year owing to the native productivity. In contrast to the situation in flowing waters, the sedimentation rate can be inferred with the aid of suitable sample collection (see Chapter 7).

LITERATURE

2.1

Ermittlung der Art und Herkunft der Sedimente im Rhein und in seinen Nebenflüssen im Bereich der Bundesrepublik (1968). Bericht 1967 für die "Länderarbeitsgemeinschaft Wasser", Koblenz.

Schwebstoffe und Schlammablagerungen in Bundeswasserstraßen (1978). Jahresbericht 1977 der Bundesanstalt für Gewässerkunde, Koblenz.

Hinrich, H. (1971). Schwebstoffgehalt und Schwebstofffracht der Haupt- und einiger Nebenflüsse in der Bundesrepublik Deutschland. Dtsch. Gewässerkd. Mitt. **15**, H. 5, 113–129.

Deutsches Gewässerkundliches Jahrbuch, Rheingebiet u. a.

Bogardi, I. (1956/57). Über die Zunahme und Abnahme der Schwebstoffgehalte in den Flüssen mit der Änderung des Abflusses, Die Wasserwirtschaft **47**, 59.

Das Neckarhochwasser vom 13.–17. Januar 1955 in der Strecke Plochingen–Mannheim (1955). Bericht der Wasser- und Schiffahrtsdirektion Stuttgart.

Van Andel (1950). Mineralogisch-petrographische Analyse der Rheinsedimente, Dissertation.

Hellmann, H., Bruns F. J. (1968). Die chemische Zusammensetzung der ungelösten Stoffe des Rheins, Dtsch. Gewässerkd. Mitt. **12**, H. 6, 162–166.

Hellmann, H. (1968). Die Belastung des Rheins mit ungelösten organischen Stoffen im Jahre 1967, Dtsch. Gewässerkd. Mitt. **12**, H. 2, 39–43.

Schmidt-Ries, H. (1958). Limnologische Untersuchungen des Rheinstromes. Bd. II, Forschungsberichte des Wirtschafts- und Verkehrsministeriums Nordrhein-Westfalen Nr. 508.

Lauterborn, R. (1905). Die Ergebnisse einer biologischen Probeuntersuchung des Rheins. Arbeiten aus dem Kaiserlichen Gesundheitsamt, Berlin.

Marson, M. (1908–1910). Bericht über die Ergebnisse der ersten im Oktober 1905 und Mai 1906 ausgeführten biologischen Untersuchung des Rheins auf der Strecke Weisenau–Mainz bis Koblenz–Niederwerth, Hefte der Kgl. Prüfungsanstalt für Wasser und Abwasser, Bd. 28.

2.2

Jahresbericht 1977 der Bundesanstalt für Gewässerkunde wie unter 2.1.

Ermittlung der Herkunft der Schlammablagerungen im Neckar, Gutachten Nlb/ 340.7/3819, Koblenz 1972.

Ermittlung der Herkunft der Schlammablgerungen im Untermain, Gutachten Nlb/ 445.2/7781, Koblenz 1974.

3

Water temperature and illumination

In the analysis of flowing waters the discharge is a parameter of essential importance. For those substances which are not only dissolved in the water but also bound to the suspended solids, the suspended solids transport must also be quantified. As a rule the concentration of dissolved solids shows a different relationship to the discharge from that shown by the suspended solids, as will be explained in Part II with the aid of specific examples.

Besides the water and suspended solids transport, and particularly where organic substances are concerned (hydrocarbons, detergents, phenols, etc.), the water temperature constitutes a third parameter. The temperature also controls the extent and rate of denitrification and the conversion of dissolved phosphate, or in short all metabolic and self-purification processes. As an example the seasonal temperature profile for the River Neckar at Lauffen in 1970 may be reproduced (Fig. 3.1). In addition the mean monthly discharge values are indicated on the graph.

The formation of phytoplankton, which alters the concentration of nutrients (N, P) and encroaches on the CO_3^{2-} and Ca^{2+} budgets, depends on the intensity of illumination as well as on the temperature of the water.

From the point of view of the suspended solids content the phytoplankton production is of major importance. In the Rhine at Koblenz during the summer months, for example, it may reach 3 mg/l or more, representing 7–15% of the total suspended solids. In eutrophic lakes it may be possible at times to obtain values of 10 mg/l.

The interaction of water temperature, level of illumination, turbidity of the water, oxygen production of the phytoplankton and dissolved oxygen content may be described with reference to Fig. 3.2.

Initially, between 1 and 7 August the water temperature fell (for reasons not apparent here) from 25 to 21°C. The cause lay in the effects of rainfall associated with a thunderstorm. The water flowrate increased from 1300 to 1650 m^3/s, while the turbidity due to entrained solids had a negative effect on the level of illumination in the water body.

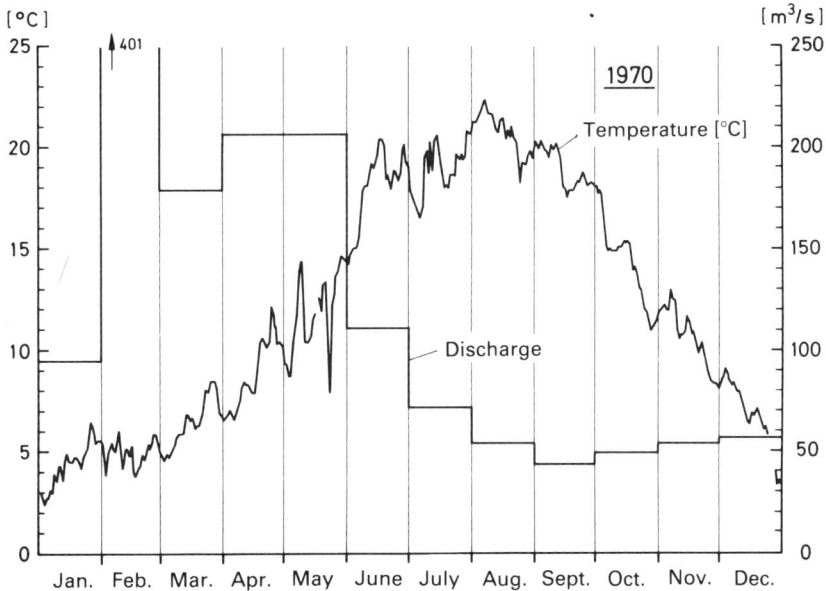

Fig. 3.1 — Water temperature and discharge in the Neckar at Lauffen, 1970.

As oxygen-depleting substances were simultaneously introduced, the dissolved oxygen level fell from 8 to 4.8 mg/l. The O_2 production of the phytoplankton is affected on the one hand by the diurnal mean value and on the other by the maximum and minimum daily readings. A good level of illumination implies specially good daytime/night-time curves, such as may be observed from 15 August. The less favourable conditions of illumination between 3 and 9 August are accordingly reflected in the smaller amplitude of variation of the oxygen content.

Due to die-off of the phytoplankton in the autumn the water body is subject to 'secondary' pollution. As particularly large quantities of oxygen are then required, the minimum dissolved O_2 value frequently occurs not in midsummer during periods of high water temperature and low flow, but in the autumn, when the water temperatures are moderate and the light and heat supply are no longer adequate for the assimilation process.

Owing to the phytoplankton productivity, distribution patterns for trace contaminants may be substantially affected. A measure of the phytoplankton productivity can be deduced from the so-called oxygen production potential (Fig. 3.3).

LITERATURE

Knöpp, H. (1960). Untersuchungen über das Sauerstoff-produktions-Potential von Flüssen. Schweizerische Z. Hydrologie XXII, Fasc. 1, 152–166.

Knöpp, H. (1959). Über die Rolle des Phytoplanktons im Sauerstoffaushalt von Flüssen. Dtsch. Gewässerkd. Mitt. 3, H. 6.

O₂ [mg/l]

Temperatur [°C]

——— Daily means

·········· Maximum/-minimum daily values

– – – 100% oxygen saturation

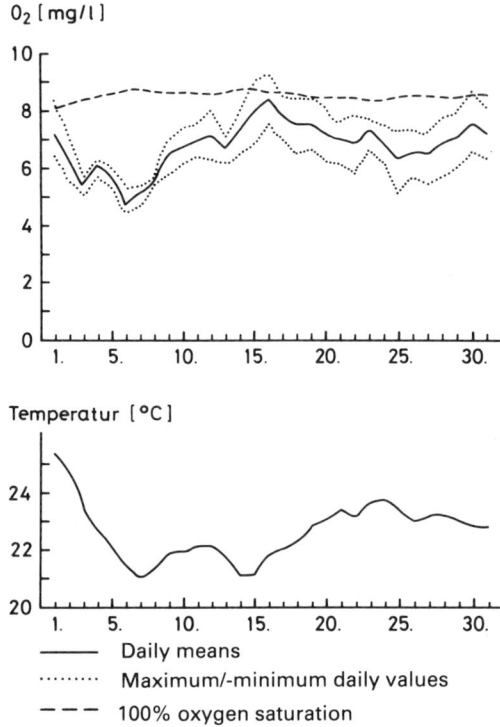

Fig. 3.2 — Oxygen content and water temperature in the Rhine at Koblenz in August 1983.

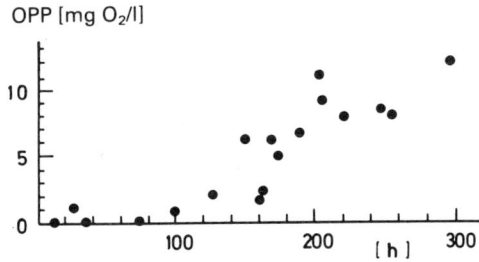

OPP [mg O₂/l]

[h]

Fig. 3.3 — Relation between oxygen production potential (OPP) and duration of sunshine
preceding the measurement (monthly mean) (from Knöpp, 1960).

Elster, H. J. (1969). Absolute and relative assimilation rates in relation to phyto-
plankton populations in Goldman, C. R. (Hrsg.) Primary Productivity in
Aquatic Environments, Mem. Int. Ital. Hydrobiol. 18 Suppl., University of
California Press, Berkeley, pp. 77–103.

Müller, D. (1977). Welchen Anteil hat das Phytoplankton am Schwebstoffgehalt von Oberflächengewässern? Dtsch. Gewässerkd. Mitt. **21**, H. 1, 1–5.

Wassertemperaturen s. Deutsches Gewässerkundliches Jahrbuch.

Wassertemperaturen des Rheins bei Koblenz s. Gewässerkundlicher Überblick, hrsg. von der Bundesanstalt für Gewässerkunde (erscheint monatlich).

4

Anthropogenic effects

4.1 NATURAL DISCHARGE AND SEWAGE INPUTS

If one surveys the rivers of the German Federal Republic, it will hardly be possible to find a site where there is no evidence of the presence of sewage, and perhaps also in a good many cases, of effluent of industrial origin. Exceptions consist mainly of springs and the headwaters of small streams.

Where potential abstraction points differ, however, within a very wide margin is in the ratio of the natural flow to the sewage input. As the daily average composition and flowrate of municipal sewage for particular localities are remarkably constant, however, it is possible to make predictions even for the longer term, which are based on numerical data applicable over an entire river basin or even throughout the whole country. For industrial effluents, however, some site-specific knowledge is of overriding importance. A large factory with an effluent discharge rate of 2 m^3/s to a receiving stream with a mean daily flowrate of 2000 m^3/s exerts an appreciably greater demand on the analyst than the situation where the receiving stream has a natural discharge of only 150 m^3/s averaged over the course of a year.

The calculation of mean values for the discharge provides a first point of reference regarding possible dilution factors. In the context of pollution abatement one normally bases one's calculations on the least favourable case, and hence for example on MNQ values, for moderately low flows. Other comparative figures are the NQ_9 or the NQ_{20} values. The NQ_{20} value is the flow which is exceeded on all but 20 days of the year; it is derived from the annual data sources for mean daily flow (Table 4.4). An impression of the discharge behaviour of the largest German rivers (apart from the Main and the Neckar) is given by Table 4.1. The annual input to coastal waters in the north-west of the Federal Republic is accordingly of the order of 100×10^9 m^3. This total includes about 10% from sewage effluents, or 5% from effluent of domestic or small manufacturing origin (Table 4.2). To what extent ground and surface waters are utilised can be inferred from Table 4.3 which gives data concerning their relative proportions for 1960–1974.

On the whole effluent from domestic or small trade premises is dilutes from 1 to 20 times on a volumetric basis. In practice, when attempting to estimate the effects of

Table 4.1 Discharge of various West German rivers

River	Gauging station	Period of observations	Mean annual total from annual series $\times 10^9 \, m^3$	1984 annual total $\times 10^9 \, m^3$	1984 relative to annual series (%)
Rhine	Maxau	1921–1980	38.37	40.02	104
	Kaub	1936–1980	50.14	53.19	106
	Rees	1936–1980	70.36	75.51	107
Moselle	Cochem	1936–1980	9.39	12.04	128
Ems	Rheine	1941–1980	1.10	1.31	120
Weser	Porta	1941–1980	5.64	6.39	113
Elbe	Neu-Darchau	1926–1980	22.64	17.66	78

Table 4.2 Quantity and origin of effluents discharged via public sewerage systems to natural waters

Calendar year	1957	1963	1969	1985
Total volume x $10^9 \, m^3$	3.60	4.90	6.40	9.70
Groundwater and small streams	0.72	0.69	0.83	1.10
Domestic and trade effluents	1.62	2.40	3.20	5.35
Industrial effluents	1.26	1.81	2.37	3.25

Table 4.3 Structural trends in water supply

Calendar year	1960	1965	1970	1974
Total volume $\times 10^9 \, m^3$	3.34	3.69	4.39	4.66
True groundwater and spring water	64.7%	63.4%	67.3%	69%
Bank filtrate and/or recharged groundwater	28.7%	28.9%	24.8%	22.4%
Lakes and reservoirs	5.0%	6.0%	6.5%	7.5%
River water	1.6%	1.7%	1.4%	1.1%

effluent discharges to the receiving stream, such a ratio is adopted. Other typical calculations are based on a total population figure for the Republic of 60 million, for whom the total available volume per annum of surface water is $100 \times 10^9 \, \mathrm{m}^3$, which is equivalent to 0.053 l per head per second (4.5 m^3 per head per day). The reciprocal value of 1/0.053, or 19, implies that for every 19 inhabitants a flow of one litre per second is available for drainage purposes.

On this basis the various river basins can be characterised and pollution loading factors calculated. Eckoldt (1962) carried out the calculation for the Neckar (Fig. 4.1) on the basis of the NQ_{20} value. At this flowrate, a part-flow of l/s is available to every 104 inhabitants of the catchment area (Table 4.4). The

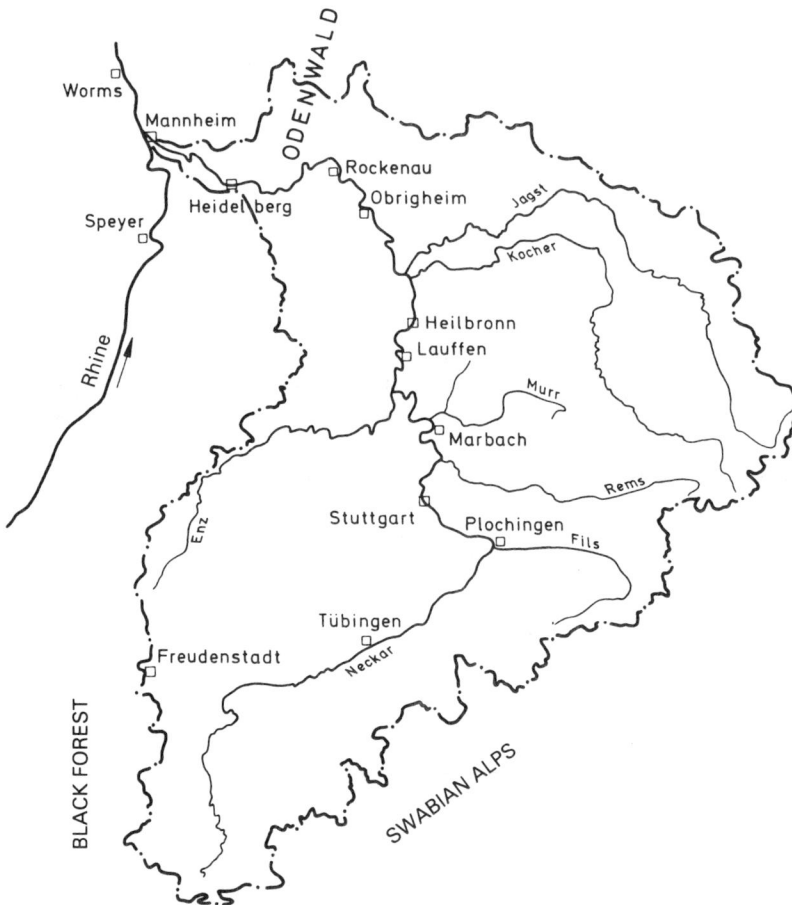

Fig. 4.1 — Map of the Neckar catchment.

middle reaches of the Neckar around the Stuttgart area can thus be regarded as especially critical in the light of their discharge characteristics. The pollution level is

Table 4.4 Characterisation of the optical effluent pollution in a catchment with reference to the Neckar (from Eckoldt, 1962)

Catchment	F Surface area (km^2)	E Pop. $\times 1000$	E/F Pop. density	NQ_{20} $(m^3/s$ $\times 1000)$	$f =$ E/NQ_{20}
Neckar above R. Fils	3295	734	223	10	73
Fils catchment	706	208	295	2	104
Neckar from R. Fils to R. Enz	1618	1234	763	—	—
Enz catchment	2223	484	218	7	69
Neckar from R. Enz to R. Kocher	671	209	311	—	—
Kocher and Jagst catchments	3826	405	106	7	58
Neckar from R. Jagst to confluence with Rhine	1627	384	236	—	—
Total	13966	3658	262	37	99

particularly critical for MNQ flows which are calculated from the 1921–1970 data at the Plochingen gauging station as 9.8 m^3/s. The lowest value recorded during this period was only 3.7 m^3/s and the sewage effluent discharged by the central sewage works for Stuttgart at Mühlhausen alone amounted to 3 m^3/s. The dilution factor in the receiving stream accordingly amounted to only 1:3.

Surface water analysis in the sense indicated at the commencement must take such factors into account — frequently at the time of sampling, but in any case when making an assessment of the results.

4.2 TRANSPORT OF NATURAL AND SEWAGE-DERIVED SUSPENDED SOLIDS

The annual suspended solids load of the Rhine is of the order of 3–4 million tonnes, that of the Neckar and the Moselle around 0.2–0.5 million tonnes (Table 2.2). By far the largest proportion of this total doubtless consists of eroded material. By and large the suspended solids concentration increases with the erosive force of the water, i.e. with the discharge.

Suspended solids of anthropogenic origin occupy a quantitatively minor position, if one considers the effect of combined effluents from domestic and small trade premises. One distinguishes primarily between combined and separate sewerage networks. In the separate system only the runoff due to rain falling on paved surfaces, such as streets, pavements, footpaths and roofs, is discharged directly to the receiving stream. This storm sewage contains principally particulate fallout from the atmosphere which for large cities can amount to 1000 kg/ha per annum and for country districts 500 kg/ha. Added to this are the products of abrasion of tyres and road surfaces, plant debris and garden soil wash-out.

The combined system continually collects the sewage of domestic and trade origin, but while the separate system allows discharges to be made directly to a receiving stream, with the combined system the sewage and runoff are transported entirely to the sewage treatment works. During rainfall events, an amount up to about double the dry weather flow usually goes to the treatment plant, the rest going directly to the receiving stream.

If one now considers the situation from the point of view of the receiving stream, then one has firstly the input of effluent suspended solids for dry weather flow conditions. As long as the sewage receives full secondary treatment, one can reckon on an annual load of 150 kg/ha of filterable solids. For a large city like Stuttgart with 150 km^2 of paved surfaces, the calculated annual load is 2250 tonnes.

On top of the dry weather solids load, there is a further input during rainfall; in the case of separate sewer systems this amounts to *ca.* 2000 kg/ha per annum. For combined systems the figure is not so large. Certainly the combined sewage and stormwater which does not reach the treatment plant will contribute additional suspended solids, and also the added hydraulic loading on the sewage works may increase the output from this source owing to the shorter retention time, but nevertheless the overall picture appears slightly more favourable in this case than for the separate system. By incorporating stormwater retention tanks the situation can be improved still further. Exact numerical values of widespread validity are not feasible nor even necessary.

If one assumes, in the case of the large city referred to, an annual contribution of 5000 t, then this is of the right order of magnitude. This quantity of suspended solids will broadly correspond to a population of 1 million. In the catchment area of the Neckar, with 3.6 million inhabitants (Table 4.4), one may therefore expect an annual solids load of 18 000 t. This compares with a total average figure for suspended solids transport of 240 000 t per annum.

If one considers lastly the purely industrial input, e.g. of sand, $CaSO_4$, silica, etc., then one can reckon that about 10–15% of the total annual solids load in the rivers of the German Federal Republic originate from sewage and effluent.

These figures do not include those materials is suspension such as bacteria and algae which are derived from the dissolved constituents of sewage effluents, which may add a further few per cent to the total.

In contrast to the behaviour of eroded material, the concentration of effluent-derived suspended solids does not increase with the discharge. Rather, it can be expected to decrease due to dilution. A critical situation arises, however, in the event of sudden short rainstorms (thunderstorms) during otherwise dry spells with low flows. A high level of pollution then enters the receiving stream via the stormwater overflows in the sewerage network (Fig. 4.2). Rainfall of such short duration leads to only a negligible increase in the discharge, but gives rise to high suspended solids contents, which are even higher for small receiving streams and densely populated catchments.

Suspended solids of anthropogenic origin therefore occur in their highest

Suspended solids [g/m^3]

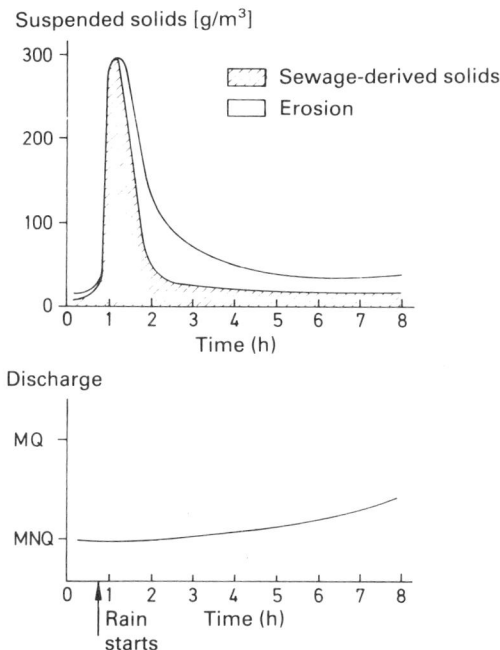

Fig. 4.2 — Composition of the suspended solids entrained in the flow following a sudden rainstorm.

percentage at times of low flow (Fig. 4.3). This is not only of importance for the analyst, but also for the water body, as it may be the cause of critical situations developing. Erosion- and effluent-derived inputs of suspended solids differ in their chemical composition, in particular in their degreee of contamination with certain trace substances. Both atmospheric dust and the solids discharge in the effluent from sewage treatment works are as a rule highly contaminated, in which respect they differ markedly from the so-called background or eroded material.

Deposits of suspended solids (sediment) form principally during periods of low water (Fig. 2.6), and therefore contain particularly large proportions of the suspended solids fractions of anthropogenic origin (Fig. 4.4).

If one takes these factors into account then one will take special care not only in respect of the acquisition of the sample, but also in the interpretation of the results, and the more so where one is required to 'scale-up'. With regard to sediments and sludges, it is generally true that the nature and method of sample removal will exert a more or less pronounced limiting influence on the intended interpretation of the results.

LITERATURE

4.1

Deutsches Gewässerkundliches Jahrbuch.
Eckoldt, M. (1962). Die Beurteilung der Abwassertbelastung eines Stromgebietes

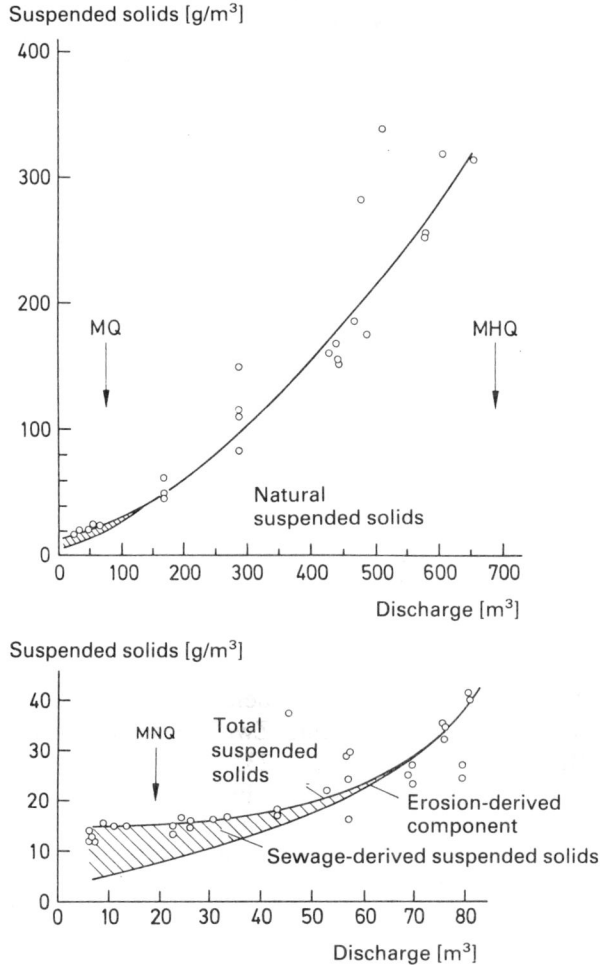

Fig. 4.3 — Proposition of suspended solids of anthropogenic origin to total suspended solids load, as a function of discharge for the Neckar.

nach den Einwohnerzahlen mit Anwendung auf das Rheingebiet, Dtsch. Gewässerd. Mitt. 6, 131–135.

Abwasser, Anfall, Behandlung und Beseitigung in Gemeinden und Industriebetrieben in der Bundesrepublik Deutschland (1972) Umweltschutz Heft 18, Hrsg. Bundesministerium des Innern.

Battelle-Institut (1973). Wasserbedarfsentwicklung in Industrie, Haushalten und Gewerbe, wasser luft betrieb **17**, 145–149.

Bundesministerium für Wirtschaft, Bonn (1984). Jahresbericht 1983 über die Wasserwirtschaft, Wasser Boden 6/7, 290–291.

4.2

Krauth, K. (1970). Der Abfluß und die Verschmutzung des Abflusses in Mischkana-

Suspended solids [mg/l]

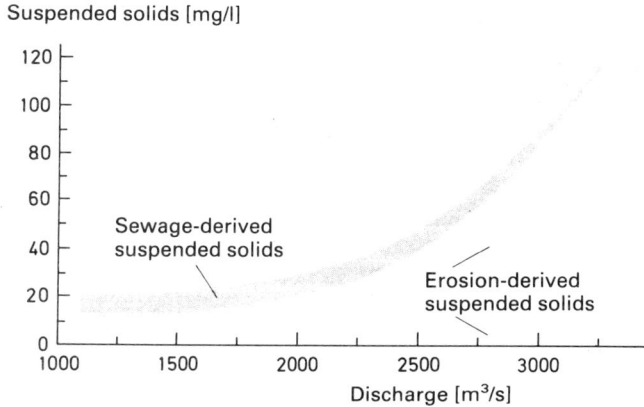

Fig. 4.4 — Proportion of suspended solids of anthropogenic origin to total suspended solids concentration, as a function of discharge, for the Middle Rhine.

lisationen bei Regen, Stuttgarter Berichte zur Siedlungswasserwirtschaft, Bd. 45, Kommissionsverlag R. Oldenbourg, München.

Pecher, R. (1974). Der jährliche Regenwasserabfluß von bebauten Gebieten und seine Verschmutzung, Korrespondenz Abwasser **21**, 113–119.

Lautrich, R. (1971). Die Schmutzwasserbelastung des Vorfluters (SBV) durch Regenüberläufe (RÜ) mit und ohne Zwischenschaltung von Regenüberlaufbecken (RÜB), Wasser Boden **8**, 234–235.

Brunner, P. G., Gewässerverschmutzung infolge Regenwasserentlastungen aus Trenn- und Mischkanalisationen. Bericht für die LAWA-Arbeitsgruppe "Normalwerte für Abwasserreinigungsverfahren", vorgelegt 8.12.72 in Essen.

Hellmann, H. (1975). Anhaltspunkte für die Berechnung der Belastung von Vorflutern durch absetzbare Stoffe aus Kanalisationen, Dtsch. Gewässerkd. Mitt. **19**, 51–54.

Ermittlung der Herkunft der Schlammablagerungen im Neckar, Gutachten N1b/340.7/3819, Koblenz 1972.

Hellmann, H. (1971). Untersuchungen über den Beitrag von Abwässern an der Schlammbildung in Bundeswasserstraßen, Z. Binnenschiffahrt Wasserstraßen, **11**, 427–431.

Hellmann, H. (1972). Herkunft der Sinkstoffablagerungen in Gewässern. 2. Mitteilung: Überlegungen und Ergebnisse aus der Sicht der Abwassertecknik, Dtsch. Gewässerkd. Mitt. **16**, 137–141.

Brunner, P. G. (1977). Straßen als Ursachen der Verschmutzung von Regenwasserabflüssen — Ein Überblick über den Stand der Forschung, Wasserwirtschaft **67**, 98–101.

Grunewald, A. (1972). Die Belastung der Vorfluter mit organischen Abfallstoffen, Korrespondenz, Abwasser **19**, 289–291.

Part II

Analysis

5

Sampling

If one reflects on the information presented in Chapters 1–4, it will be seen that the taking of a representative sample from a flowing watercourse is no easy task. The water flow and the suspended solids content are constantly changing, as also are the temperature and illumination conditions. In fact one ought to devote much more space to the taking of the sample for this topic to be correctly and factually treated. However, only some aspects will be considered in more detail, while from time to time futher mention may be made in the succeeding pages, in Chapter 7–13.

One can state that the taking and the pretreatment of the sample must be pre-arranged having regard to the substances to be determined, the origin of the sample itself and the particular problem in hand.

If, for example, the object is to determine the load of a heavy metal such as cadmium at a particular point for a total period of 1 year, then quite different conditions apply from the situation where it is desired to obtain a longitudinal profile or to pinpoint the location of an effluent discharge. In the author's experience it is advisable, for those parameters which are partly in the dissolved and partly in the suspended state, to separate water and sediments *before* performing the analysis; however, other methods are often advocated. A particular problem arises in the case of sludges or sediments. Water–suspended solids–sediment: these terms imply a triple subdivision of the sampling process.

5.1 WATER SAMPLES

Sampling method
Water samples may be obtained using automatic samplers on or in the water, e.g. on bridge piers and in power stations. For spot samples specially designed apparatus of plastic or glass construction is often employed, enabling samples to be obtained at various depths. For surface samples a bucket may be sufficient.

Sampling time and place
For certain 'simple' parameters — among these are pH, temperature, conductivity and dissolved oxygen — continuous measurement during both the day and the night is possible. For comparative studies in the course of the year spot samples should all

be taken at the same time of day. If one desires to assess the water quality with reference to an agreed system of classification, then sampling is performed during dry weather, if possible under conditions of low flow (MNQ). The same conditions are appropriate for determination of longitudinal profiles. On no account should such measurements be performed during wet weather (see Figs 1.1 and 1.3).

For estimation of trends, identical discharge characteristics for several successive years are desirable (e.g. MQ). On the other hand, the study of discharge-related changes in the relevant materials concentration may be undertaken at a particular point.

The question of the homogeneity of the water body at the point in question is very important under certain circumstances and should be checked beforehand if necessary. Substances which occur almost exclusively in the dissolved state in water are Cl^-, NO_3^-, NH_4^+, Na^+ K^+.

Both Ca^{2+} and Mg^{2+}, and PO_4^{3+} to an even greater degree, exist in equilibrium with the biochemical processes occurring in the water. Most heavy metals and trace organic contaminants, as well as detergents, humic acids and bulk parameters such as BOD and COD, are as a rule somehow related to the suspended solids transport. A total determination of soluble plus insoluble components is frequently only possible with difficulty, e.g. as a consequence of incomplete extraction and matrix effects. For this reason the water and suspended solids phases should be separated prior to analysis. Organic trace constituents are often partially absorbed onto filter paper, hence separation should be performed by centrifuging or by settling of the sample into the two fractions.

5.2 SUSPENDED SOLIDS SAMPLES

Method of sampling

A variety of different suspended solids collectors have been developed, which entail differing water volumes. For the analysis of the solids, the particulate fraction in a litre of water is usually insufficient. Frequently 10 or more litres must be obtained. For surface samples several buckets or vessels made of plastic may be filled with up to 100 litres of water. In our experience these can be left to stand overnight and the following day the supernatant liquid can be drawn off down to a residual volume of 0.25 l (for a 10 l bucket). After transferring the solids adhering to the bottom and walls of the bucket into a beaker, the settling process is repeated.

For obtaining larger samples for the purpose of determining temporal or spatial average values, as well as perhaps for the recovery of samples from different depths, a continuous centrifuge may be employed. The water pumped up from the river bed in excess is fed in part through the centrifuge running at 35 000 rev/min (feedrate *ca.* 180 l/h). The volume of clear water leaving the centrifuge is measured. The solids collected in the machine can be flushed out into a measuring cylinder after stopping the flow, and considered relative to the volume of clarified liquid obtained.

Less suitable for this task are sampling devices situated either on land or in the river, through which water is allowed to flow continously. As a consequence of insufficient settling times, the solids are to some extent fractionated and the coarser but correspondingly less polluted fraction is retained in the collector. For stations on land beside the river, this process may also take place to some extent in the connecting pipework.

Occasionally clay suspensions are not inclined to settle. In this case an artificial flocculation by means of Al_3^+ ions may be necessary.

The water-containing suspended material is frequently freeze-dried or air-dried at 80–105°C. In so doing the possibility of losses with respect to certain substances cannot be excluded. A preliminary test would appear advisable. Sometimes the water remaining after centrifuging may be chemically combined by triturating with Na_2SO_4. Where organic substances are to be determined, this is followed by a solvent extraction in a Soxhlet apparatus.

Some indication of the distribution of suspended solids with depth beneath the surface is given in Table 5.1 while typical fluctuations in the surface layer are recorded in Fig. 5.1.

Fluctuations in the content of suspended solids are often accompanied by a change in their chemical composition (Table 5.2). Such changes are generally small in the horizontal plane but large in the vertical direction. Invariably the solids content is greatest close to the bed of the river while the specific organic pollution conversely increases near the surface.

Sampling time and place

For the sampling time the same observations are applicable as those already indicated for water samples. Particular problems can arise in flowing waters of low transport capacity, e.g. in artificially regulated rivers at low water. As the entrainment action in the upper layers of impoundments can be very slight, sampling in this region is of little benefit; the transport of suspended solids takes place just above the bed of the river. In the tailwater one obtains only the 'residual solids fraction' which has not been deposited as sediment behind the weir. At higher values of discharge, sampling in both the headwater and tailwater regions is admissible (Fig. 2.6).

For small watercourses and for discharges of only a few cubic metres per second and below, the suspended solids transport may reflect the influence of local inputs. Then it is frequently advisable to take into account a certain diurnal pattern. This is true of the determination of the annual phosphate loads for medium-sized streams.

5.3 SAMPLES OF SEDIMENT OR SLUDGE

The standard DIN 4045 Part 1, Item 3.5.1 refers to 'sediments as deposited constituents', while according to Item 4.3.1, the term sludge is understood to mean the 'moisture-containing solids separable from effluents'. In the analysis of natural waters both terms are encountered. Deposits of high moisture content and a high concentration of organic matter (≥15% by weight) are usually referred to as sludge. The dividing line between sediment and sludge, however, is indistinct.

In free-flowing waters deposits of finely-divided solids are not present, but sand and gravel are found instead. These are as a rule of little interest to the analyst. In controlled waters such as harbours, however, between retaining structures and within all types of impoundment, fine-grained or sludge-like deposits are formed.

Table 5.1 — Suspended solids distribution in different vertical planes of the channel cross-section

Place	Date	Discharge (m³/s)	Depth (cm)	Suspended solids content (mg/l)[a]
Breisach l.h.s.	26.6.1967	1797	50	44.3
			240	46.4
			375	49.2
			560	49.2
Mainz middle	15.8.1967	1640	50	21.8
			200	32.1
			350	35.3
			430	39.2
Koblenz l.h.s.	17.8.1967	1720	50	29.2
			150	32.7
			370	35.5
Koblenz middle	16.8.1967	1670	50	26.5
			150	30.5
			250	31.8
			370	36.7
Koblenz r.h.s.	17.8.1967	1720	50	29.5
			150	32.1
			370	37.2
Kaiserswerth r.h.s.	12.5.1967	2060	50	26.8
			130	30.6
			480	33.3
			525	32.2
Frontier r.h.s.	12.5.1967	2070	50	36.4
			130	37.4
			340	49.6
			535	50.8

[a]After drying at 105°C.

Frequently one may observe all stages in the transition from sand to clay or silt fractions within a very small compass (Fig. 2.8).

The study of such solids is only meaningful within the scope of a precise definition

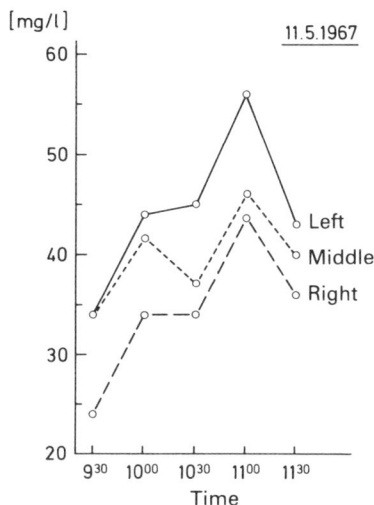

Fig. 5.1 — Fluctuations in the suspended solids content of the Rhine at the frontier with
the Netherlands — surficial water layer.

of the problem, and with the benefit of a good site plan, e.g. a map of navigation routes and industrial premises. If one desires to locate the supposedly most heavily polluted zone within fairly close limits, then the loss of ignition and water content are useful indicators. In an impounded reach, this zone is situation close to the weir and up against the mole which divides between the weir and the lock channel (Fig. 2.8). One can obtain a sample with the aid of a small scoop or a hollow plunger (Fig. 5.2).

In investigating dredge spoil a representative measurement may be requested. For the reasons stated above this can hardly be achieved by taking an underwater sample. The task can be performed more readily by sampling the dredged material from the chute or from the drainage field. Even then, however, quite considerable uncertainties may arise for scaled-up estimates.

In stationary waters such as inland lakes the incoming material is similarly fractionated. The finest fractions travel the greatest distance and sink to the greatest depth (see Fig. 7.14, 250 m). In favourable circumstances a depth profile may be obtained with the aid of a hollow plunger which will allow not only annual rates of deposition, but also particular pollution events and trends to be identified and assigned to certain dates.

The procedure of fractionating sediment samples according to particle size prior to analysis and then employing the fraction $<20~\mu$m or $<6~\mu$m for assessing the level of pollution is quite legitimate and meaningful within closely defined comparative situations. The extent to which such a finely-divided fraction may be typical of a particular sampling point and the pollution introduced by a point-source discharger must, however, be determined by means of further investigation involving the hydraulic regime, among other things. In general it is not admissible to scale up the values obtained from analysis of sediments in flowing waters to give annual fluxes at a given measuring station.

Table 5.2 — Chemical composition of suspended solids in the Rhine at various points: A — Mainz, km 498; B — Kaub, km 547; C — Mainz, km 498 at different depths

Place	Date and time	Discharge (m³/s)		SiO₂ (%)	Al₂O₃ (%)	CaO (%)	MgO (%)	Fe₂O₃ (%)	CO₂ (%)	Loss on ignition (%)	Sampling depth (cm)
A	3.4.67 10.30–14.00	1830	left	45.4	8.6	8.5	2.7	4.9	5.5	22.2	80
			middle	45.8	9.7	6.7	2.4	5.4	4.5	23.1	80
			right	47.9	10.7	3.3	2.0	6.0	3.2	26.9	80
B	4.4.67 11.00–15.00	1900	left	43.8	10.0	6.8	3.0	4.7	4.9	22.5	80
			middle	45.0	10.0	8.8	2.8	4.6	4.4	23.0	80
			right	44.1	10.8	6.8	2.5	5.2	3.8	24.5	80
C	15.8.67	1650	left	44.9	5.1	10.5	2.1	4.0	9.1	18.2	150
			middle	40.3	4.7	10.6	2.9	4.0	8.8	20.0	150
				44.3	6.5	11.3	2.5	3.6	8.9	18.8	430
			right	34.9	5.9	7.2	2.3	8.2	5.3	32.5	150
				40.7	5.1	8.8	2.0	4.6	7.6	26.0	450

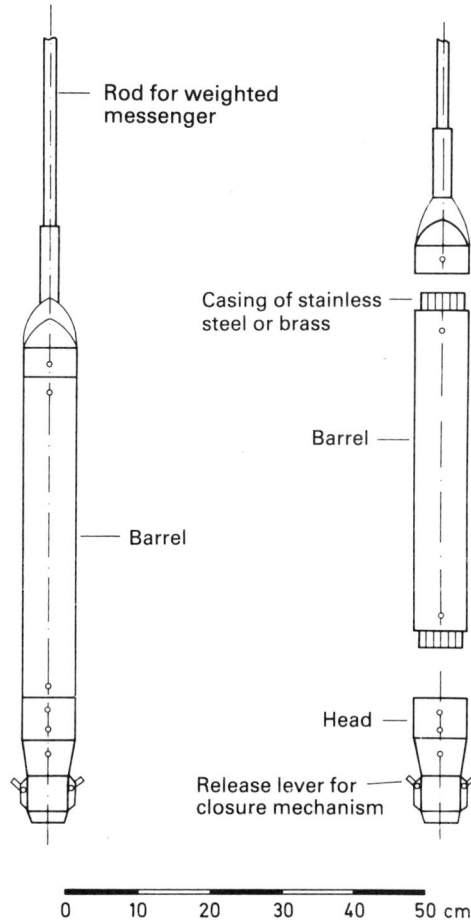

Fig. 5.2 — Coring tool for sampling sediment and sludge deposits for vertical profile studies.

LITERATURE

Kalweit, H. (1976). Die Station zur kontinuierlichen Messung der Rheinwasserbeschaffenheit in Mainz, Wasser Boden **11,** 283–287.

Klein, K., Krems, W., Wellner, B. (1976). Betriebserfahrungen und Verbesserungsvorschläge zum "Elektrodenspülsystem Koblenz", Dtsch. Gewässerkd. Mitt. **20,** 79–84.

Hellmann, H. (1974). Zur Inhomogenität von Flußwasserparametern, Dtsch. Gewässerkd. Mittl. **18,** 23.

Brinkmann, F. J. J., Zoetmann, B. C. J., Piet, G. J. (1973). Inhomogene Flußwasserqualität — Automatische Messung. Dtsch. Gewässerkd. Mitt. **17,** 159–166.

DIN 38414 Teil 11: Hinweise zur Probenahme von Sedimenten aus stehenden and fließenden Gewässern (Vorschlag Stand 13.11.1984).

Ermittlung der Art und Herkunft der Sedimente im Rhein und in seinen Nebenflüs-

sen im Beriech der Bundesrepublik (1968). Zu G/340.7/1100, Bundesanstalt für
    Gewässerkunde.
Treunert, E., Wilhelms, A., Bernhardt, H. (1974). Einfluß der Probenentnahme-
    Häufigkeit auf die Ermittlung der Jahres-Phosphor-Frachtwerte mittlerer
    Bäche, Hydrochemische und hydrogeologische Mitt. **1,** 175–198.
Fischer, M. (1972). Meßstationen für die kontinuierliche Überwachung der Wasser-
    güte, Wasser- and Energiewirtschaft **64,** 215.
Stadler, D., Schomaker, K. (1977). Ein Glaskugelschöpfer zur kontaminations-
    freien Entnahme von Seewasser unter der Oberfläche, Dtsch. Hydrogr. Z. **30,**
    20–25.
Der Rhein. Schiffahrts- und Industriekarte des Rheins, Binnenschiffahrtsverlag
    GmbH Duisburg-Ruhrort, MST 1:15000.

6

Classical parameters

6.1 PRELIMINARY REMARKS

By 'classical parameters' we understand those substances which have been employed for the characterisation of natural waters and their water quality status for many years. They are generally fairly easily determined, using relatively simple methods, such as those specified in the 'German Standard Methods' and in the successor DIN series. From time to time improvements have been made to some of these methods.

For the classical quality parameters we are usually concerned with concentrations in excess of 1 mg/l. With the exception of 'chloride' and dissolved oxygen in the past, they do not hit the headlines of the daily press so often as recent trace contaminants. In recent times both phosphate and nitrate have become topical subjects.

The timeless importance of the classical parameters for a first assessment of water quality is not to be doubted. Classical methods of analysis cannot be replaced by modern methods for trace-substance analysis in this context. Even where there are no particular problems for the analyst in connection with the analysis itself, nevertheless they frequently arise in the course of interpretation of the results. He may disclose them in an instant in response to a question about water quality regardless of distinctions affecting the actual time and spatial significance of the data. Not infrequently the analyst is subject to embarrassment as a result. Archival data cannot be used *per se* to provide up-to-date information. The widespread inability to give a precise answer on the spot can be overcome by promptly applied methods of data assessment. Classical parameters for lightly polluted waters provide a good introduction to this subject.

6.2 CATIONS AND ANIONS PARTICULARLY IN LIGHTLY POLLUTED WATERS

On the Upper Rhine, at the point where the Untersee or westernmost portion of Lake Constance narrows to become a free-flowing river, lies the beautiful little Swiss town of Stein am Rhein. At this point also is situated the first sampling point of the

International Commission for Protection of the Rhine against Pollution.† As the discharge averaged over the year is very large (over 300 m³/s) in relation to the population in the catchment area (0.9 million), and also because Lake Constance functions largely as a preliminary settling basin for materials of anthropogenic origin, the measurements at this point provide an indication of 'background' levels for various constituents.

The results may be presented either as a chronological series or as a function of discharge. If one takes the chloride concentration of spot samples as an example, then during 1975 these fluctuated in the range 3–8 mg/l (Fig. 6.1) (the first measurement in September is probably erroneous). The corresponding mass trans-

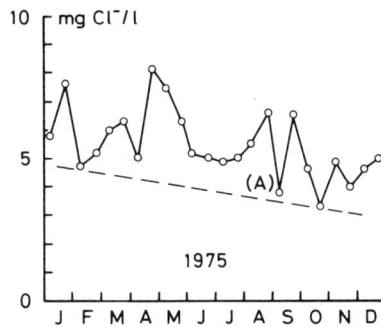

Fig. 6.1 — Chloride concentration in the Rhine at Stein — seasonal variation.

port rate increased from spring to summer by a factor of two- to five-fold and showed a maximum during August (Fig. 6.2). The reason for this can be found in changes in

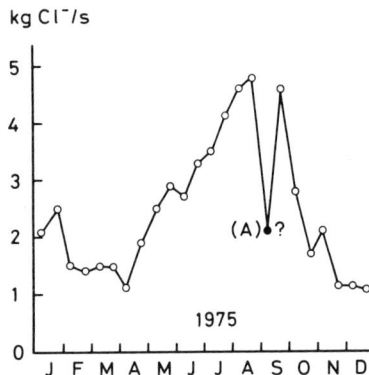

Fig. 6.2 — Chloride mass transport rate in the Rhine at Stein — seasonal variation.

† Data for Chapter 6 are largely derived from tables of analytical data published by the Commission.

the discharge, which can be quickly discerned from a graph correlating the mass flow and the discharge (Fig. 6.3). The linear increase in the chloride transport rate with the summertime snowmelt in the mountains and the resulting increase in discharge carries an unmistakable message. One is evidently dealing here with pollution of predominantly natural origin in the waters of the Rhine. The variability in the concentrations determined by analysis should not be given undue prominence. On a larger canvas (Fig. 6.4) these are insignificant, e.g. when compared with the anthropogenic chloride inputs and their fluctuation, such as indicated by results for the Rhine at Koblenz.

If the transport/discharge relationship alters significantly over a longer period of time, such as for chloride between 1965 and 1975 (Fig. 6.5), one may suspect human interference as the cause, whether from enhanced sewage inputs or maybe as a result of hydraulic engineering measures affecting the discharge regime.

If one considers the sulphate ion, which emanates like chloride from geogenic sources (Fig. 6.6), in relation to the discharge, then once again a linear increase in the mass transport rate as a function of water flowrate is obtained. The relationship is quite rigorous. A similar increase in the load to that for chloride between 1965 and 1975 is not discernible. At Stein am Rhein the sulphate concentration is already greater than that of chloride by a factor ten. The further uptake downstream to Koblenz is, however, within the same order of magnitude. One may infer from this that the anthropogenic sulphate pollution is of the same order as that of geogenic origin.

Temperature and illumination are entirely independent factors, each of which may influence the results. Thus if the Ca^{2+} concentration is correlated simply with the water temperature then this may serve as an expedient. The level of natural illumination definitely plays a dominant role in determining the concentration of dissolved calcium (Fig. 6.8). The phenomenon of biogenic decalcification, which is addressed in Fig. 6.8, will be discussed more fully in Section 6.5.

The quantitatively most important constituents of the Upper Rhine can be seen from Table 6.1 to be Ca^{2+} and HCO_3^-, followed by Mg^{2+} and SO_4^{2-}. The ions Na^+ and Cl^- are much less important, while K^+ and NO_3^- occur in even smaller concentration. The value for orthophosphate lies below 0.1 mg/l. Increased flow-rates change these concentrations only slightly, if at all. In this context, the chemical composition of the Upper Rhine is relatively easy to review and interpret.

Anthropogenic effects are of lesser importance than geogenic factors. The materials concentrations are approximately constant, which implies an increasing transport rate with increased discharge. Nevertheless biogenic processes do have an impact on the materials balance, such that a change in the equilibrium between dissolved and undissolved materials (Ca^{2+}/HCO_3^-) may be brought about.

6.3 NITRATE: DISCHARGE BEHAVIOUR AND LONGITUDINAL PROFILE

The nitrate ion is, from an entirely analytical viewpoint, a 'conservative' parameter which like the chloride and sulphate ions remains constant in natural waters. In reality nitrate is involved in the dynamic interactions of the metabolic processes in natural waters. In contrast to the 'dynamic' phosphate fraction, however, the availability may also change and under certain conditions molecular nitrogen (N_2)

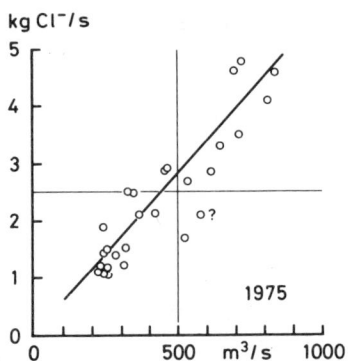

Fig. 6.3 — Correlation between chloride transport and discharge in the Rhine at Stein.

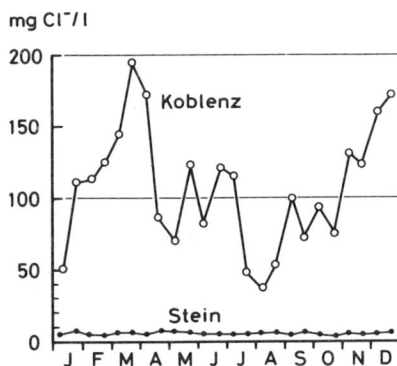

Fig. 6.4 — Chloride concentrations in the Rhine at Stein and Koblenz in 1975 — seasonal variation.

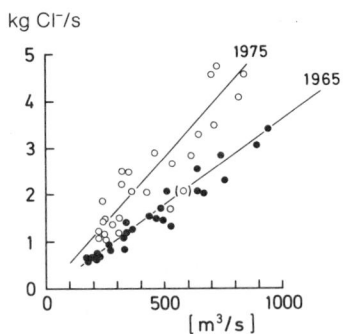

Fig. 6.5 — Correlation between chloride transport and discharge in the Rhine at Stein for 1965 and 1975.

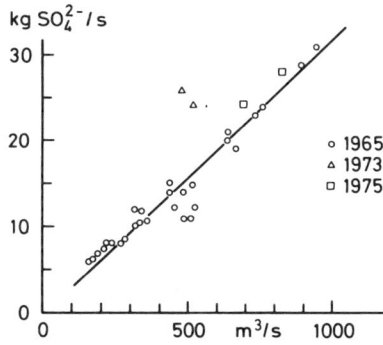

Fig. 6.6 — Correlation between sulphate transport and discharge in the Rhine at Stein, 1965–1975.

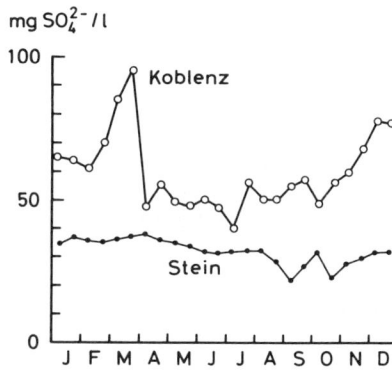

Fig. 6.7 — Sulphate concentrations in the Rhine at Stein and Koblenz during 1965.

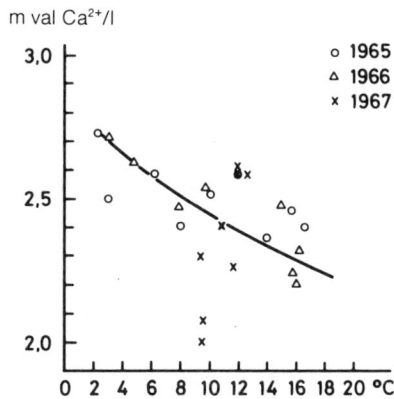

Fig. 6.8 — Correlation between Ca^{2+} concentration and water temperature, Stein, 1965–1967.

Table 6.1 — Total analyses at Stein am Rhein, 1965 (data in meq/l)

Date	Discharge (m^3/s)	Temp. (°C)	Ca^{2+}	Mg^{2+}	Na^+	K^+	HCO_3^-	SO_4^{2-}	Cl^-	NO_3^-
31 Jan.	205	3.1	2.50	0.70	0.16	0.03	2.40	0.75	0.10	0.05
15 Mar.	165	2.4	2.73	0.41	0.19	0.03	2.36	0.77	0.12	0.04
27 Apr.	439	6.1	2.59	0.53	0.16	0.03	2.36	0.73	0.11	0.04
10 Jun.	739	10.1	2.51	0.80	0.14	0.03	2.48	0.65	0.11	0.04
24 Jul.	760	15.9	2.46	0.71	0.14	0.05	2.36	0.66	0.08	0.04
30 Aug.	462	16.6	2.40	0.66	0.13	0.04	2.36	0.56	0.09	0.04
10 Oct.	490	14.0	2.36	0.48	0.13	0.04	2.29	0.58	0.08	0.03
21 Nov.	262	8.1	2.40	0.99	0.16	0.03	2.33	0.67	0.10	0.04

may be formed. If one first of all ignores these dynamic changes, the relation between the NO_3^- transport and discharge is very informative (Fig. 6.9). It exhibits a roughly linear increase of mass flowrate as the discharge increases. The scatter of individual values is somewhat greater than that for Cl^- and SO_4^{2-} in the Rhine at Stein. From the relatively unambiguous relationship one might infer an exclusively natural source of nitrate contamination for the middle Rhine. Also the graphical picture (Fig. 6.9) shows hardly any difference between the five-year periods for 1965–1970 and 1971–1975.

That some other factors are having a measurable effect on the particular concentrations is clear from a plot of the nitrate concentration against water temperature (Fig. 6.10): increasing water temperatures are clearly associated with lower nitrate contents. This double dependence on water temperature and flowrate naturally leads to some scatter in the daily mass transport rates. For this reason to quote an exact transport figure for a 12-month period is hardly possible. Still more problematical, for the same reasons, is an analysis of trends in the case where hydrological and biochemical factors are not equally weighted or cannot be so treated. A further insight into this topic is provided by the yearly extreme and mean values cited in Table 6.2. The concentration maxima are invariably recorded in winter, the minima in summer. In the course of the dynamic metabolic sequence of events, ammonia (which is not referred to here) is converted into the nitrate/nitrite form (nitrification) while on the other hand the process of denitrification takes place simultaneously with liberation of N_2; the final outcome is fixation of NH_3 in protein of bacteria and also phytoplankton occurring within the context of primary productivity. Even the last-named process may be responsible during the summer for mopping up several milligrammes of NH_3 per litre.

According to season, the nitrate concentration, as may be evident from the Table, can fall by as much as 50% from winter to summer. The fact that the nitrate pollution of the Middle Rhine, and possibly elsewhere as well, has noticeably increased between 1965 and 1979 cannot be ignored. The minimum values increased from 2.55 to 7.2 mg/l during this period, the averages from 5.74 to 12.0 mg/l and the maxima from 8.0 to 19.4 mg/l.

Despite this obvious result, the causes should not merely be ascribed to enhanced sewage-related inputs without further examination. The transport/discharge behaviour suggests other possible sources of nitrate pollution than domestic sewage, such as agriculture.

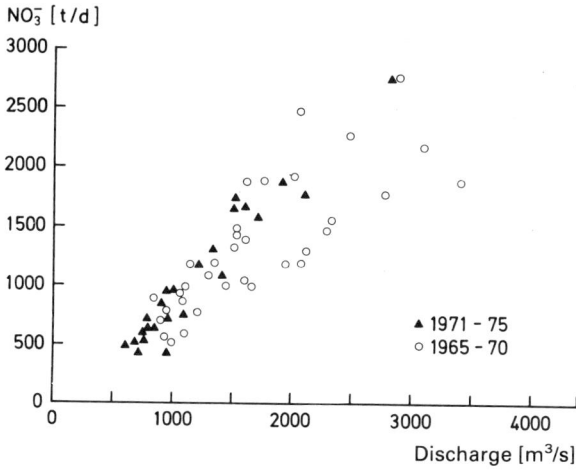

Fig. 6.9 — Correlation between nitrate transport and discharge in the Rhine at Kaub.

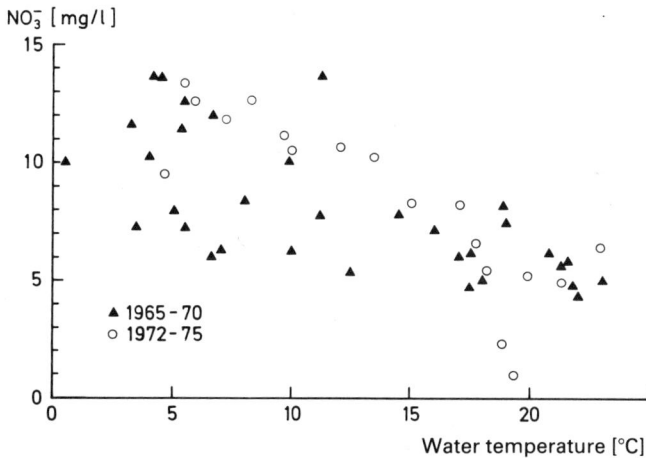

Fig. 6.10 — Correlation between NO_3^- concentration and water temperature in the Rhine at Kaub.

The picture is amplified by a longitudinal profile. Such a profile was determined between 26 and 30 September 1983 in the direction of flow, under constant discharge conditions (MQ) (Fig. 6.11). In so doing the nitrate concentration was shown to rise by a factor of 8 between Constance (1.9 mg/l) and the national frontier (15.8 mg/l). Peak values were recorded below the mouths of certain tributaries (Main, Wupper). If one regards the value obtained at Constance as representing the natural level of background contamination, then the anthropogenic increase is certainly substantial.

Table 6.2 — Nitrate content of the Rhine at Kaub

Year	Date	Min. (mg/l)	Max. (mg/l)	Average ($n = 26$)	Discharge (m^3/s)
1965	18 Sep.	2.55		5.74	2460
	18 Jan.		8.00		2330
1966	11 Jun.	3.40		5.62	1930
	19 Jan.		9.95		1580
1968	31 Jul.	5.4		9.1	1450
	9 Nov.		13.8		1650
1975	29 Jul.	5.4		9.4	2100
	23 Feb.		13.4		1510
1977	11 Jun.	7.2		10.7	1660
	3 Feb.		18.1		1920
1979	25 Jun.	7.2		12.0	2290
	9 Mar.		19.4		2120

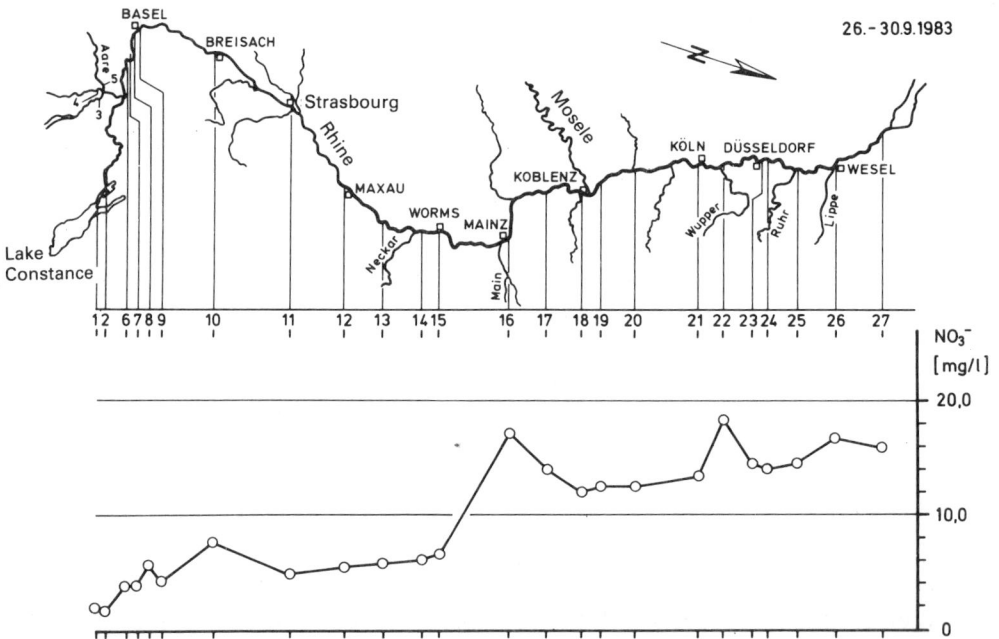

Fig. 6.11 — Nitrate concentrations in the Rhine — longitudinal profile, 26–30 September 1983.

Measurements of the nitrate contents of spring water, water close to the source and rainwater show concentrations ranging from 2 to 18 mg NO_3^-/l (Table 6.3) These are concentrations which are not exceeded in a good many natural waters. If one considers the aspect of artificial fertiliser application in agriculture and considers the special retention capacity of the soil at the vegetative season, together with the converse of the increased wash-out of nitrate ion during the winter, then the problem reduces to one of nitrate balance estimation and evaluation of nitrate contents.

For the sake of completeness the pollution peaks should also be mentioned which can occur in cases of sustained flood flows, especially when sludge deposits with a high ammonia content become entrained and suspended in the water. Owing to the chemical reactions which set in very rapidly, the oxygen content of the water may be markedly diminished, while the nitrate content increases.

Even if nitrate has been the major object of attention during the preceding paragraphs, the ammonium ion should not be assigned an unimportant place. In sewage-polluted watercourses, the level of ammonia-N may be roughly equal to that of nitrate-N, and like the latter it tends to a minimum during periods of elevated temperature in the water.

6.4 PHOSPHATE: DISCHARGE BEHAVIOUR AND LONGITUDINAL PROFILES

Phosphate may be regarded among other things as an indicator of sewage effluent inputs to natural waters, especially since the natural background is characterised by values less than 0.1 mg PO_4^{3-}/l and anthropogenic inputs in connection with sewage discharges in West Germany amount to something of the order of 100 000 t per annum.

Analytically it is possible to distinguish fairly easily between three 'categories', viz. the soluble *ortho*-phosphate, the total phosphate, and the insoluble phosphate associated with the suspended solids or sediment. Besides the soluble *o*-phosphate, the dissolved phosphate fraction may also include other combined forms of phosphorus, such as combined organic phosphorus, for example.

While in springs, in streams close to the source and in rainwater the *o*-phosphate ion recedes with respect to the other combined forms of phosphorus, the *o*-PO_4^{3-} fraction usually predominates in sewage-polluted watercourses. Like nitrogen, phosphorus is subject to various permutations in its speciation as a result of the various chemical and biological processes occurring in natural waters. However, in this case, in contrast to nitrogen, the level of total-P remains constant. Under these circumstances a mass balance with regard to the origin and fate of phosphorus in natural waters, as well as trend analyses, can be performed, but only if all possible combined forms of phosphorus are taken into consideration.

Important deductions can, however, be made solely from a consideration of values for *o*-phosphate. In the Moselle at Koblenz, for example (Fig. 6.12), the *o*-PO_4^{3-} concentration during 1961 ranged from 0.1 to 0.2 mg/l irrespective of the discharge. Consequently the mass transport rate increased in proportion to the discharge. Fourteen years later, in 1975, there was evidence of a clear increase in concentration to 2 mg/l at times of low flow (Fig. 6.12). Simultaneously a characteristic discharge–concentration relation had developed in the shape of a hyperbola.

Table 6.3 — Nitrate content of non-polluted sources, streams near the source and rainwater

Location	NO$_3^-$ (mg/l)	Ref.
Kreuzstein spring — Wesergebirge	2.6–6.0	
Woodland spring — Wesergebirge	30.0–36.0	
Wümme spring, Lüneburg heath	0.5–1.8	K. Höll
Haverbeck spring, Lüneburg heath	0.3–0.8	
Fulda source — Rhön	0.8–1.5	
Sieg source — Rothaargebirge	5.0–16.0	
Lahn source — Rothaargebirge	1.3–6.2	Fed. Institute
Ochsenborn — Rothaargebirge	1.4–3.7	for the Science
Streams — Rothaargebirge	1.5–16.0	of Natural Waters
Oeser district — Harz region	11.0–21.6	(Koblenz)
Ocker district — Harz region	8.8–17.2	1976 and 1977
Black Forest springs	0.9–20.2	(unpublished
Weidelbach — Black Forest	1.0–6.8	data)
Rainwater at Koblenz 1976	1.1–1.8	
Rainwater at Koblenz 1984	2.4–5.8	as above
Snow at Koblenz 1976	3.2	

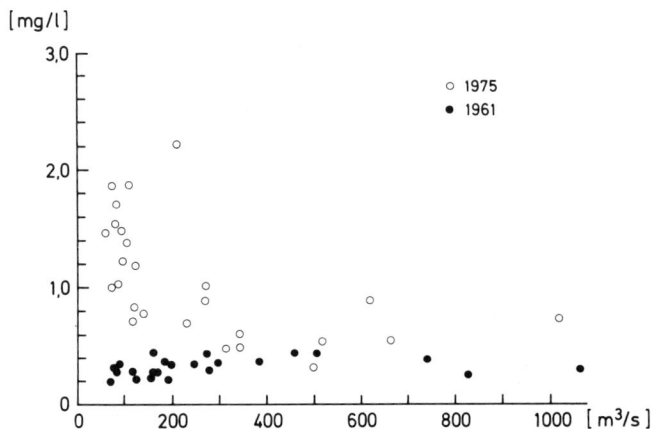

Fig. 6.12 — Orthophosphate concentration vs discharge in the Moselle at Koblenz during 1961 and 1975.

Similar discharge–concentration curves were also typical of other rivers, especially for the Middle Rhine, during this period.

If one follows the legitimate assumption that phosphate is derived chiefly from domestic wastewaters and, in the case of a sufficiently large catchment area, imposes a relatively constant pollution load on the receiving waters, then the hyperbolic concentration curve becomes intelligible: higher discharges imply a more pronounced dilution of the pollution load (Fig. 6.13).

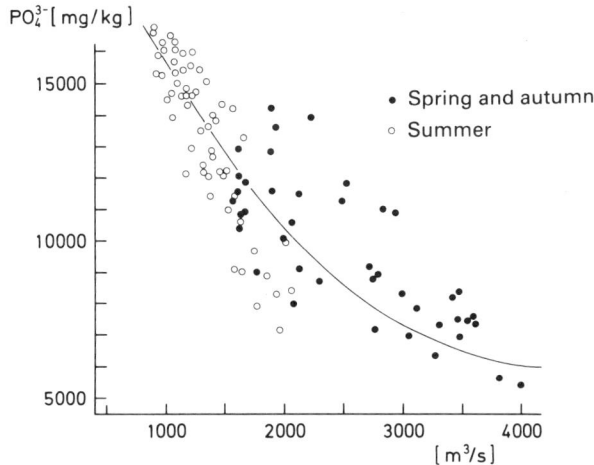

Fig. 6.13 — Phosphate contents of suspended solids in the Rhine as a function of discharge at Koblenz in 1974.

Besides the analysis of soluble phosphates, determination of total phosphate is also advisable. The hypothesis that under flood conditions relatively uncontaminated eroded material will 'dilute' the more heavily polluted sewage-derived suspended solids, is confirmed by the behaviour shown in Fig. 6.14. If, as was shown in Chapter 2, the overall suspended solids load increases with the discharge, then by the same token the transport of insoluble phosphates will likewise increase with the discharge.

The potential extent of the changes in concentration during periods of rising and falling water flowrates, caused by rainfall of limited duration, is shown by the relevant analytical data (Fig. 6.14). By and large the concentration profile exhibits the previously mentioned pear-shaped pattern which is very plainly evident in the case of the total-phosphate contents plotted as a function of discharge.

The concentration–discharge behaviour as it has emerged during the period 1960–1970 for the larger West German rivers enables the following characteristic features to be inferred. With increasing discharge there is:

— a decrease in the level of soluble orthophosphate
— a decrease in the phosphate content of suspended solids

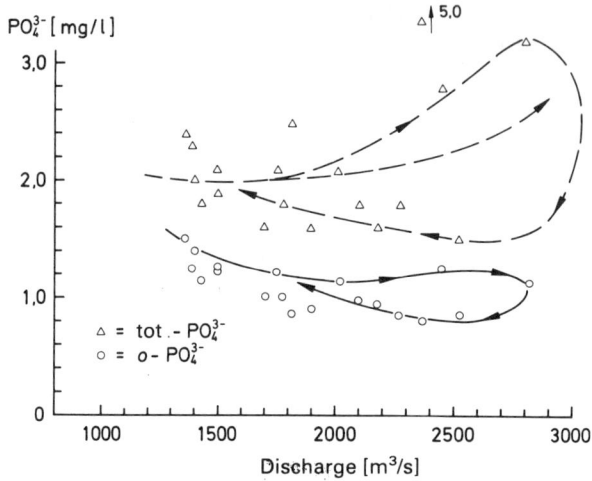

Fig. 6.14 — Concentration profile of o- and tot-PO_4^{3-} for rapidly rising and falling flows in the Rhine at Koblenz, in 1978.

— a corresponding increase in the level of total phosphate
— an increase in the mass transport of total phosphate

In simplified form the relationship for the Middle Rhine may be broadly represented by Fig. 6.15 in which the organically bound phosphate is represented by the non-shaded portion.

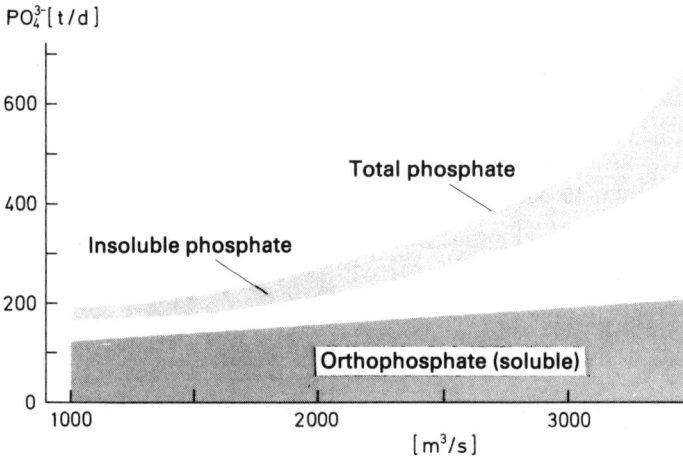

Fig. 6.15 — Heavily amplified picture of the dependence of phosphate transport on discharge, with respect to different chemical forms of phosphorus, in the Rhine at Koblenz, 1974–1976.

In smaller rivers and streams the measurements may of course turn out quite differently. Thus a diurnal profile of the PO_4^{3-} contents with peak values in the morning and evening, perhaps slightly displaced in time, may be expected where discharges from small communities without adequate sewage interception and treatment facilities are involved. In wet weather periods the much less heavily polluted suspended solids from cultivated and forested areas will become of greater significance than is the case for the major rivers.

For longitudinal profiles one may expect an increase in the phosphate content in the direction of flow. In the case of the Rhine downstream profile (Fig. 6.16), the

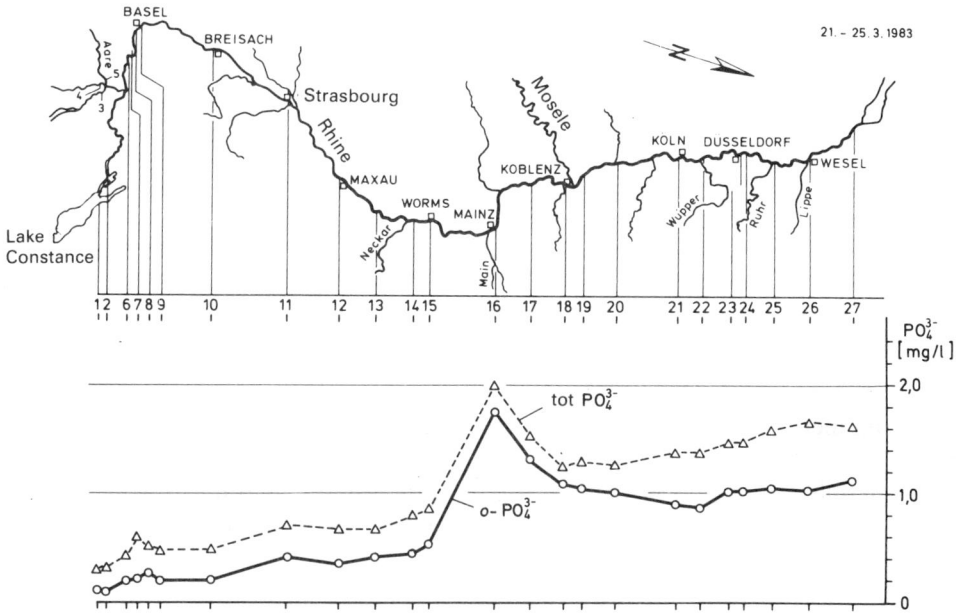

Fig. 6.16 — Increase in o- and tot-PO_4^{3-} concentrations along the length of the Rhine from Constance to Emmerich/Lobith, 21–25 March 1983.

o-PO_4^{3-} contents rise considerably while the insoluble and organically bound phosphate show smaller concentration increases under conditions of dry weather flow. The transport rates corresponding to the figures in the diagram amount to 240 t per annum for the Middle Rhine at Koblenz and 540 t per annum at the Dutch–German frontier.

Owing to the presently excessive rate of fertilisation of natural waters, no noticeable concentration changes are to be expected as a result of the summertime increase in primary (algal) productivity. For the same reasons one could expect to achieve a reduction in algal productivity only as a consequence of a vast reduction in the P-input.

As phosphate currently occupies a major role in the concepts of environmental protection — it is necessary only to mention the tertiary treatment stage for sewage and the Maximum Phosphate Ordinance for detergents — the correct analytical determination of the phosphate† and the consideration of the mass transport–dis-

† Analytical procedures for o-PO_4^{3-}: spectrophotometric method with addition of ammonium molybdate; for total phosphate: similarly to o-PO_4^{3-} after extraction with sulphuric acid and potassium persulphate.

charge relationships is a centrally important task of the analytical chemist. It is just as indispensable for the underpinning of advanced sewage-treatment techniques as for political counselling on environmental issues.

6.5 OXYGEN

An example of the particularly multifaceted importance of a classical parameter is provided by the oxygen content of natural waters. In the assessment of water quality it occupies, as always, a place of central importance. Its concentration in source and pristine flowing waters is in the first place temperature-dependent. Over and above this is the turbulence of the water and the associated equilibration with the atmosphere. These natural factors bring about a seasonal and diurnal profile for the content of dissolved oxygen.

Sewage inputs as a rule contribute degradable organic substances and also inorganic compounds such as ammonium ions, owing to which the dissolved oxygen content undergoes a further fall depending on the water temperature and the saturation value.

Furthermore the O_2 production of the phytoplankton, primarily during the summer months, has an appreciable impact on the oxygen budget of natural waters. This 'biogenic aeration' (after Knöpp) can be on a par with the physical aeration. Besides the oxygen input during the daytime there is a process of oxygen depletion occurring at night. Even if the production outweighs the depletion, critical situations may nevertheless arise in natural waters. Thus in the Moselle daytime/night-time fluctuations of 13 mg O_2/l may be recorded. These, however, are extreme values, of the kind which do not occur in the free-flowing Rhine.

If one considers the situation peculiar to the Middle Rhine at Koblenz then one is dealing here with an observation point which is favourably affected by the turbulent region of the Rhine Gorge. The next major sewage effluent discharges occur only 90 km further downstream. Despite the intensive physical aeration, during the first half of November 1971 under conditions of very low flow (NNQ) only 1–3 mg O_2/l was recorded. An increase in the water flow to MNQ values was followed by a rise in the daytime/night-time values for dissolved oxygen to 4–6.5 mg/l (Fig. 6.17).

Similar circumstances arose in November 1972. Under average low flow conditions dissolved oxygen levels of 2–4 mg O_2/l were recorded. From the 16th, however, the discharge increased by several hundred cubic metres per second to average values (MHQ) which made possible a return to more favourable oxygen concentrations of up to 8 mg/l.

Noticeable on the one hand is the major importance of the magnitude of the discharge, and on the other the indirect effect on the measured values of the massive concentration of degradable substances, as shown by a comparison between the recorded O_2-content and the theoretical saturation values. Following the nadir for dissolved oxygen levels on the Middle Rhine in 1969, a distinct improvement in oxygenation occurred as a consequence of the well-known efforts to upgrade sewage treatment plants On 20th October 1974 the water flowrate had fallen to the MQ level of 1300 m^3/s, and the oxygen content remained at 3–5 mg/l. After heavy rainfall the discharge increased to the MHQ figure of 3000 m^3/s and the O_2-concentration again reached a peak of 8 mg/l (Fig. 6.18).

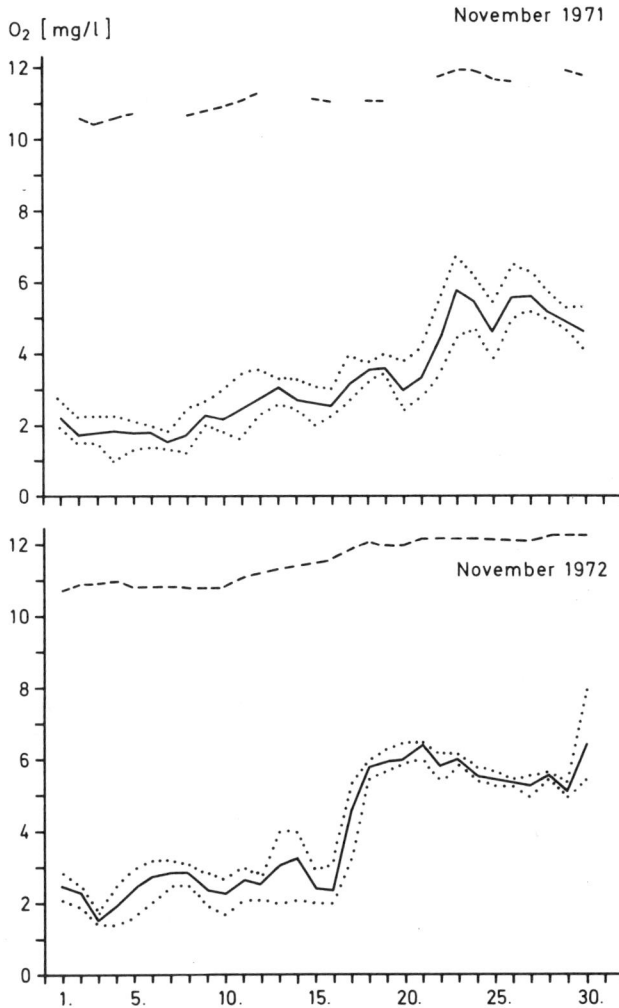

Fig. 6.17 — Oxygen contents in the Rhine at Koblenz in November 1971 and 1972. Upper line
= oxygen saturation curve.

The success of the sewage treatment measures could not be overlooked in September 1980, when the daily average reached 9 mg/l for conditions of average water flow, and was thus close to the saturation value. Although the dry weather discharge fell towards the end of the month to the MNQ value, the oxygen content did not fall below a value of 5 mg/l. The diurnal fluctuation of the oxygen content during the low water periods was clearly discernible.

A very extended consideration of the daily O_2-content and its diurnal profile in conjunction with the discharge values is essential for an assessment of the water quality situation as well as of the influence of sewage treatment facilities in the relevant catchment area. Thus the information which one obtains from the annual

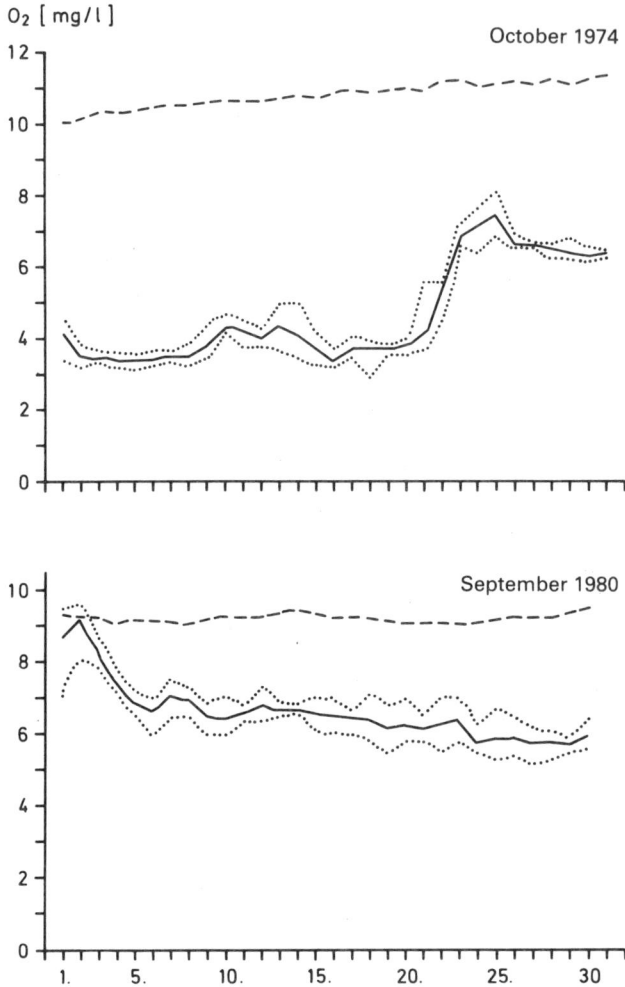

Fig. 6.18 — Oxygen contents in the Rhine at Koblenz in October 1974 and September 1980.

mean values (Fig. 6.19) is far from being of comparable benefit. Any consideration of mean values without any reference to the hydrological conditions will only provide the desired information with great difficulty. In conclusion some indication of the change in the pH value associated with the biochemical processes should not be omitted, which should equally be related to the oxygen regime.

6.6 Ca^{2+}/CO_2 SEASONAL PATTERN

The large amounts of suspended solids which are transported downstream by the Rhine daily and yearly are still only about 5% of the quantity of dissolved solids present in the water. In addition to 1 million tonnes of suspended solids there are about 20 million tonnes of dissolved solids transported annually. The particulate

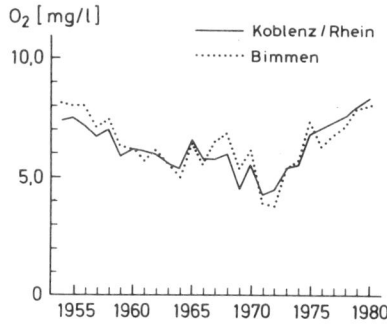

Fig. 6.19 — Annual mean values of oxygen content in the Rhine at Koblenz and Bimmen/
Lobith, 1955–1980.

solids swim in a kind of 'chemical field' and it is to be expected that certain interactions will occur between the dissolved and particulate fractions. These may be of a basically biochemical or purely chemical nature, without disregarding the fact that in a particular case the boundary between the two may be difficult to define. Chemical and biological processes have their own governing principles, which to some degree overlap.

Thus the process of self-purification in conjunction with the decomposition and transformation of organic matter affects the materials balances of silica, phosphorus and oxygen as well as the equilibrium system $CO_2/Ca(HCO_3)_2/CaCO_3$. Merely due to the formation of diatoms in the Lower Rhine, 0.4 g SiO_2 can be removed from every cubic metre of water and combined in their skeletal and capsular structure. This is equal to about 4% of the dissolved SiO_2 content of the water. For the oxygen budget, the aeration capacity of the phytoplankton in the Rhine can equal or even exceed that estimated for the physical aeration process. Thus on the one hand the nutrient supply leads to an increased biomass production, which reduces the dissolved solids fraction, while on the other hand the biochemical transformation processes to some extent convert insoluble organic matter into soluble products such as, for example, humic acids.

A purely chemical process is illustrated by the solubilisation reaction:

$$CaCO_3 + CO_2 + H_2O \rightarrow Ca(HCO_3)_2$$

Superimposed on this, however, especially in the summer, is the photosynthesis reaction of the phytoplankton. Formation of algal biomass proceeds according to the following equation (Stumm and Morgan, 1970):

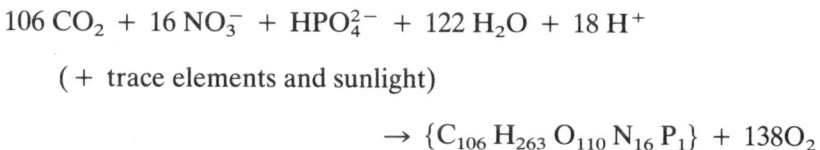

$$106\,CO_2 + 16\,NO_3^- + HPO_4^{2-} + 122\,H_2O + 18\,H^+$$

(+ trace elements and sunlight)

$$\rightarrow \{C_{106}\,H_{263}\,O_{110}\,N_{16}\,P_1\} + 138\,O_2$$

Photosynthesis leads visibly to the multiplication of algae, but also (invisibly to the human eye) to a disturbance of the Ca^{2+}/CO_2 equilibrium.

As the dissolved carbonic acid in the pH range 7–8 largely exists in the form of bicarbonate (HCO_3^-) ions, this is subject to the effect of biogenic activity to a particularly marked degree.

According to Fig. 6.20 a concentration minimum for dissolved Ca^{2+} and HCO_3^- is characteristic of the summer months. This applies not only to relatively unpolluted

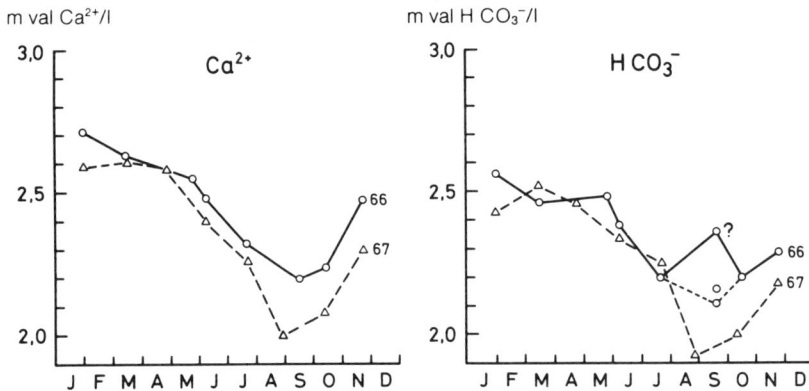

Fig. 6.20 — Seasonal profile of Ca^{2+} and HCO_3^- concentrations in the Rhine at Stein, 1966/67.

but also to heavily polluted stretches of water (Fig. 6.21).

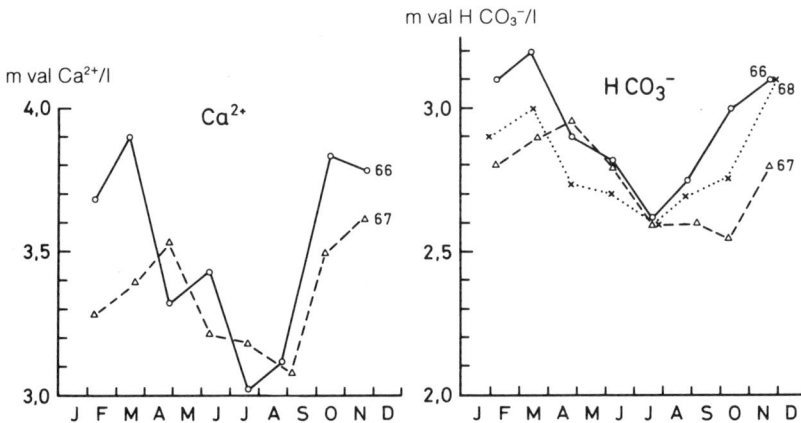

Fig. 6.21 — Seasonal profile of Ca^{2+} and HCO_3^- concentrations in the Rhine at Braubach, 1966/1967.

A biogenic precipitation of 0.2 mol Ca^{2+}/l such as was observed at Stein am Rhein during the summer indicates the formation of 20 mg $CaCO_3$ which in the insoluble form enhances the particulate solids fraction. At higher rates of phytoplankton production the effect, already considerable, becomes even greater. The changes in chemical composition of the suspended solids fraction, in both the organic and inorganic constituents, which is apparent to the analyst, are thus first and foremost of a seasonal nature, and inherently associated with the materials conversion involved in the process of photosynthesis.

In the next chapter, on heavy metal analyses, it will be shown that biogenic decalcification contributes to the precipitation of manganese and zinc and presumably other trace elements in addition. Thereby the process achieves a prominent position in the context of heavy metal contamination of suspended solids and sediments. The study of the materials distribution between dissolved and insoluble phases is the more effective the more the hydrobiological processes are taken into account. The peculiar nature of the materials cycle in lakes may be alluded to.

LITERATURE

6.2

Internationale Kommission zum Schutze des Rheins gegen Verunreinigung. Zahlentafeln der physikalisch-chemischen Untersuchungen des Rheins sowie der Mosel/Koblenz., Ab 1965.

Haberer, K. (1961). Untersuchungen über die anorganischen Inhaltsstoffe des Oberrheins, Vom Wasser **28**, 33–43.

6.3

Höll, K. (1964). Langjährige chemische Untersuchungen unbeeinflußter Quellen und Quellbäche, Vom Wasser **31**, 26–42.

Bundesanstalt für Gewässerkunde (1976 und 1977). Gewässerkundliche Untersuchungen über die Dynamik des Umsatzes von Phosphat, Nitrat und Borat im Rhein, Koblenz.

Bucksteeg, W. (1966). Lastpläne uber den Eintrag von Phosphor- und Stickstoffverbindungen als Beurteilungsgrundlage für die Wahl der Maßnahmen zur Eindämmung der Gewässereutrophierung, Abwassertechnik **17**, Nr. 3, I–II; Nr. 4, III–IV.

Schwille, F. (1973). Die chemischen Zusammenhänge zwischen Oberflächenwasser und Grundwasser im Moseltal zwischen Trier und Koblenz, Besondere Mitteilungen zum Deutschen Gewässerkundlichen Jahrbuch Nr. 38, Koblenz.

Sturz, O. (1969). Betrachtungen zur chemischen Beschaffenheit des Mains in den Abflußjahren 1958–1967, Dtsch. Gewässerkd. Mitt. **13**, 66–73.

Müller, D., Kirchesch, V. (1981). Nitrifikation in Fließgewässern. Bedeutung-Messung-Berechnung, Vom Wasser **57**, 199–213.

Müller, D., Kirchesch, V. (1983). Zur Nitrifikation in Mosel und Donau, Arch. Hydrobiol. Suppl. **68**, 63–76.

6.4

Treunert, E., Wilhelms, A., Bernhardt, H. (1974). Einfluß der Probenentnahme-Häufigkeit auf die Ermittlung der Jahres-Phosphor-Frachtwerte mittlerer Bäche, Hydrochemische und hydrogeologische Mitteilungen **1**, 175–198.

Hamm, A. (1983). Stoff-Fracht-Schwankungen in einem alpinen Fluß, Vom Wasser **60**, 123–145.

Hellmann, H. (1977). Zur Phosphatbelastung des Rheins in gewässerkundlicher Darstellung, Gas-Wasserfach, Wasser-Abwasser **118**, 259–263.

Zobrist, J., Davis, J. S., Hegi, H.-R. (1977). Charakterisierung des chemischen Zustandes von Fließgewässern, Gas–Wasser–Abwasser **57**, 402–415.

Brümmer, G., Lichtfuß, R. (1978). Phosphorgehalte und bindungsformen in den Sedimenten von Elbe, Trave, Eider und Schwentine, Naturwissenschaften **65**, 527–531.

Müller, G., Nagel, U., Jhan Purba (1978). Gelöste Borund Phosphorverbindungen in Neckar und Elsenz: Herkunft, Bilanzierung. Umweltrelevanz. Chem. Ztg. **102**, 169–178.

Mańczak, H. (1968). Über die Auswertung von Gewässergüteuntersuchungen, Vom Wasser **35**, 237–265.

Hamm, A. (1977) Uraschen und Quellen der Phosphorbelastung unserer Oberflächengewässer, Wasser Berlin **177**, Kongressvorträge, S. 186–202.

Wernet, J., Ebert, J. (1977). Über die Relevanz von Tripolyphosphat als Komplexbildner im Wasser und Oberflächenwasser, Z. Wasser Abwasser-Forsch. **1**.

Bundesanstalt für Gewässerkunde (1979) Stellungnahme zur Auswirkung der geplanten Phosphathöchstmengenverordnung auf die stauregulierte Mosel, Gutachten für das Bundesministerium des Innern.

6.5

Hofius, K. (1971). Der natürliche Sauerstoffgehalt eines Fließgewässers, Beitr. Hydrologie **1**, 1–10.

Schmassmann, H. (1955). Die Stoffhaushalts-Typen der Fließgewässer, Arch. Hydro.-biol. Suppl. XXII, 504–509.

Knöpp, H. (1959). Über die Rolle des Phytoplanktons im Sauerstoffhaushalt von Flüssen, Dtsch. Gewässerkd. Mitt. **3**, 139–147.

Wolf, P. (1973). Gewässerkundliche Zustandsaufnahme und -analyse als Grundlage für Sauerstoffhaushaltsbetrachtungen, Wasserwirtschaft **63**, 1–5.

6.6

Axt, G. (1961). Die Kohlensäure-Gleichgewichte in Theorie und Praxis, Vom Wasser **28**, 208–226.

Schmitz, W. (1961). Die fließende Welle — eine Betrachtung über die Dynamik des Energie- und Stoffhaushalts in Flüssen, Vom Wasser **28**, 11–32.

Mauz, J. (1968). Die Entwicklung der freien Kohlensäure im Vergleich zum Sauerstoff als Kriterium für den Gütezustand des Bodensees, Vom Wasser **35**, 109–136.

Stumm, W., Morgan, J. J. (1970). Aquatic Chemistry, John Wiley, New York, S. 553/554.

Ohle, W. (1934). Chemische und physikalische Untersuchungen norddeutscher Seen, Arch. f. Hydrob. **26** und weitere Arbeiten.

Elster, H. J. (1962). Stoffkreislauf und Typologie der Binnengewässer als zentrale Probleme der Limnologie, Die Naturwissenschaften **49**, 49–55 und weitere Arbeiten.

Eberle, H., Hennes, Ch., Dehnad, F. (1982). Berechnung und experimentelle Prüfung eines komplexchemischen Modells der Hauptkonstituenten des Rheinwassers, Z. Wasser Abwasser-Forsch. **15,** 217–229.

Schwoerbel, J. (1971). Einführung in die Limnologie, Gustav Fischer-Verlag, Stuttgart, S. 66–67. English translation: Handbook of Limnology, Ellis Horwood, Chichester, 1986.

Auf die Jahresberichte der Arbeitsgemeinschaft Rheinwasserwerke e.V. (ARW) sowie der Berichte über die Jahrestagungen der Internationalen Arbeitsgemeinschaft der Wasserwerke im Rheineinzugsgebiet (iawr) sei hingewiesen.

7

Trace metals

7.1 PRELIMINARY REMARKS

At the beginning of the 1960s certain new groups of compounds came to the fore in the analyses of natural waters, including hydrocarbons, pesticides and heavy metals.

The extension of classical analysis into the modern trace analysis techniques brought with it different problems. Thus, for example, in the case of heavy metals more or less expensive types of instrument were called for, such as atomic absorption spectrometers (AAS), X-ray fluorescence spectrometers, polarographs or neutron activation instruments, in place of the simple spectrophotometer of previous years. The at first fairly simple mode of operation of these instruments was later complicated by the addition of auto-sampling devices and computers.

Besides this purely instrumentation aspect the problems of sample pretreatment became very much more acute. The great importance of sample pretreatment and clean-up, in addition to that of sampling itself, has been underestimated right down to this day. Trace substances can be directly determined in water and sediment samples without serious errors ($\pm20\%$) only under very strict conditions. Numerous discrepancies in the analytical results may be ascribed to the so-called 'matrix-effect'. According to recent ring tests, not only were very substantial differences encountered between the results of different groups employing the same method (e.g. AAS) but the use of different methods (AAS, XFA, NAA) tended to produce differing results. Certainly a high level of reproducibility in the results for one particular method is no guarantee for the accuracy of the result—for more details see Doerfel (1962) and also the standard methods.

In this connection the special experience of the analyst with regard to possible sources of contamination and matrix effects has not been rated sufficiently highly. In a particular case it may be worth even more than a very expensive high-performance analytical instrument. Apart from the critical consideration of the analytical method, there are also some points of reference for a plausibility check by those not directly involved with the heavy metal analysis. They concern a comparison of individual results over a certain period of time and also for longitudinal profiles in the downstream direction. According to experience, large jumps in the concentrations of heavy metals at a particular sampling point on a large river occur only rarely. They should in any case occasion a check on the analytical procedure. In the longitudinal direction one also expects a certain regularity and probable limits for the results. This will be discussed further in Sections 7.4 and 7.5.

The importance of trace metals in surface waters has somewhat shifted 'departmentally' during the last two decades. While one first of all was chiefly concerned with the abstraction of surface waters for the production of potable supplies—still of considerable importance today—later the immobilisation of heavy metals in the sediments and suspended solids occupied the leading position. In this connection two separate problems can be distinguished: the first is the possible adverse consequences of sludge deposition on land (several million cubic metres of dredge spoil are invariably disposed of annually in the Federal Republic); the second is the possible accumulation of heavy metals in aquatic food chains and systems. A third focus of attention in connection with research on natural waters must involve the behaviour of heavy metals with regard to adsorption and remobilisation and in the presence of certain impurities of natural origin (humic acids) or artificial origin (e.g. NTA).

7.2 BRIEF DETAILS OF ANALYSIS

The determination of selected trace metals in natural waters has become a subject of very intensive study, both for many working groups in the academic world and for the official bodies responsible for water quality conservation—see literature references. The correspondingly comprehensive literature on the subject is such that it is beyond the scope of this work to undertake a detailed review of methods. However, from many years' experience it may be helpful to add a few comments.

The heavy metals are embedded in a matrix, which frequently affects the outcome of the measurement. Thus for water samples a separation may be effected by means of ion-exchange or liquid–liquid extraction. Where the enriched heavy metals are then reacted with the aid of organic complexing agents, such as sodium diethyldithiocarbamide, one obtains a precipitate which provides a uniform matrix, ideal for analysis by means of the X-ray fluorescence technique. The same applies to the combination liquid–liquid extraction and terminal analysis by AAS.

The concentrations of particular heavy metals may differ by more than a factor of ten. After enrichment from the water phase, iron and manganese predominate, followed by zinc. Mercury, together with cadmium, cobalt and nickel, are ordinarily encountered in very low concentrations. The effort required for their determination can sometimes be considerable.

The problem of analysis of sediments and suspended solids is somewhat different. An acid extraction does not always give 100% yield. This arises particularly in the determination of chromium of which a part is always derived from natural sources (see Table 7.1). Although the use of X-ray fluorescence methods for sediment and solids samples is less demanding on time and effort, it is not successful for very small amounts of, e.g., Cd and Hg. Moreover one must beware of the 'particle size effect' which cannot necessarily be equated with the matrix effect. One can control the former by careful selection of definite size fractions. Thus one may perform the analysis on fractions with particle size limits of < 0.2 mm, < 0.060 mm, or < 0.002 mm.

Table 7.1 — Average heavy metal content of igneous and sedimentary rocks from Turekian and Wedepohl (1961) quoted by Rösler and Lange (1976), mg/kg

05	Basalt	Syenite	Acidic high in Ca	rocks low in Ca	Granite, etc.	Schist	Schist & clay	Sand-stone	Carbon-aceous
Cd	0.22	0.13	0.13	0.13	0.1	0.3	0.3	0.0X	0.035
Cr	170	2	22	4.1	25	90	100	35	11
Cu	87	5	30	10	20	45	57	X	4
Hg	0.09	0.0X	0.08	0.08	0.08	0.4	0.4	0.03	0.04
Pb	6	12	15	20	20	20	20	7	9
Zn	105	130	80	60	60	95	80	15	20

The columns "Igneous rocks" span Basalt through Schist; "Sedimentary rocks" span Schist & clay through Carbonaceous.

The analysis of trace metals is certainly not a matter of routine. Although the widely used AAS method is well suited to the determination of those metals of the greatest interest in the soluble phase, at the same time the risk of contamination during sample collections and pretreatment is very high. In the analysis of seawater the difficulties for inexperienced operators may be multiplied several-fold.

Neutron activation analysis offers certain advantages regarding the accuracy of results, but does not cover all possible metals, and is also very costly in terms of equipment.

If all the different research and analytical objectives in the context of environmental protection are considered, it is hardly likely that a single method based on only one type of instrument could be regarded as satisfying the requirements of all interested parties.

Bearing in mind the understandable desire of all responsible analysts to expand further the capabilities of trace metals analysis, the danger must not be underestimated that this process will become self-perpetuating. For numerous practical purposes, an accuracy of ±20% is quite sufficient and in certain cases maybe ±50% or even 100%. The competence of the analyst then expresses itself in his ability to 'call a halt'.

Ring tests lasting several years also present a danger that the capacity of the laboratory to carry out tasks of practical importance will be eroded.

7.3 BACKGROUND AND ANTHROPOGENIC POLLUTION

The term 'background pollution' is frequently employed in the literature to mean 'pollution of a medium emanating from natural sources'. Trace metal analyses of water and suspended solids samples normally provide no indication of the ratio of the background to the anthropogenic (and hence superimposed) level of pollution. The situation differs somewhat from the conditions obtaining in radiological or isotopic research. There the cosmic β-radiation may certainly interfere with the measurement of artificially induced contamination, where it cannot be excluded by suitable screening. However, this background radiation is of very constant intensity and can thus be very easily allowed for when interpreting the results (Fig. 7.1).

For trace metals in surface waters the background component is much more difficult to determine. It appears to be just as variable as the natural heavy metal content of soils and sediments which may differ widely between regions and localities. Owing to the fluctuating input of the products of weathering under changing climatic conditions, the already variable input of heavy metals of anthropogenic origin becomes even more difficult to define accurately. As the anthropogenic fraction, just like the background, is by no means constant (Fig. 7.2) an expert interpretation of the analytical results for a particular task becomes possible only in the light of numerous measurements, previous experience and a certain attention to the hydrological circumstances, such as was described in the opening part of this volume.

A guide to the background levels for argillaceous rocks is provided by the data of Turekian and Wiedepohl. Over and above these are the so-called soil limiting values of Kloke (1977) (Table 7.2). These almost invariably allow a rough estimate to be made of the trace metal contents of suspended solids and sediments.

Since the natural suspended solids content of flowing waters increases to a greater or lesser extent with the discharge, the ratio between anthropogenic and background heavy metal contamination will change simultaneously, as shown in heavily simplified form in Fig. 7.3. Thus suspended solids loads for periods of high flow are less heavily contaminated than those during low flow periods. In favourable situations it is possible to arrive at a rough estimate of the background from consideration of the correlation between discharge and heavy metal content (see also Part III). The quantification of a flood event generally shows that the mass transport of heavy metals increases with the discharge, while the transport peak almost invariably precedes the discharge peak flow. The transport peak incorporates, depending on the catchment, a high proportion of heavily contaminated anthropogenic deposits which were built up during dry weather periods in sewerage pipes, regulated reaches of rivers and the more quiescent zones of natural waters, and which are carried along on the flood wave. In the Middle Rhine several peaks may be observed in succession, when first of all the Main, then the Neckar and finally the Upper Rhine add their contributions to the main river. Depending on the types of effluent of industrial origin in the respective catchments, separate peaks may be recorded for cadmium, lead, chromium and possibly other heavy metals.

The undissolved heavy metals behave in a manner totally different from the dissolved fraction. Because for the dissolved fraction a similar 'stockpiling' during the dry weather periods does not occur, the onset of rainfall can *de facto* only dilute the anthropogenic dissolved fraction. Peak values are therefore not to be expected. Exceptions may, however, arise in catchments where ores are mined. For very exact studies the inherent metal content of the rainwater must not be overlooked.

According to the ideas sketched out above, there are three different measuring programmes regarding heavy metal concentrations and mass flows which should concern the surface water analyst, viz. dry weather flows, incipient flood flows and flood flows.

The combination of measured data and diagrammatic illustration in Fig. 7.4 leaves the second phase, that of incipient flood flows, out of account. However, it is possible to see clearly, in the case of the Rhine gauging point at Koblenz, the basic behavioural features of the dissolved and undissolved heavy metal fractions as a

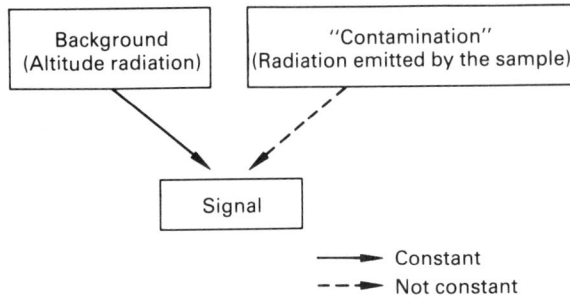

Fig. 7.1 — Make-up of the instrument reading for radiochemical measurements.

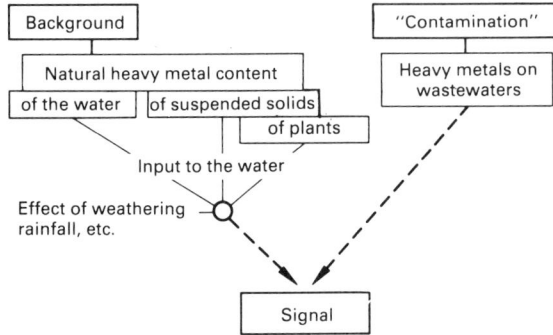

Fig. 7.2 — Make-up of the instrument reading for the determination of trace metals in natural waters.

function of discharge, with the undissolved fraction subdivided into anthropogenic and (estimated) background components.

According to our present state of knowledge (1985) the dissolved zinc for low rates of water flow is still only present in amounts of 30–40 μg/l, a visible consequence of the introduction of improved sewage treatment during the last decade. Diagrams of the type of Fig. 7.4, however, give no indication of the interactions between the dissolved and undissolved heavy metals. Whether and to what extent true equilibria with greater accumulation tendencies on the particulate suspended solids are established, must remain an open question. This applies especially to the situation for flood flows. On the other hand, special studies of the type referred to in Section 7.4 provide evidence that biochemical processes exert a massive impact on the distribution of matter.

Table 7.2 — Natural heavy metal contents in argillaceous rocks from Turekian and Wiedepohl, soil limiting values from Kloke and the limiting values for sewage sludge prescribed by the Sludge Disposal Ordinance.

	Clay Standard value (ppm)	Soil limiting value (Kloke) (ppm)	Limit for sewage sludge (ppm)
Lead	20	100	1200
Cadmium	0.3	3	30
Chromium	90	100	1200
Cobalt	19		1200
Copper	45	100	1200
Nickel	68	50	200
Mercury	0.4	2	25
Zinc	95	300	3000

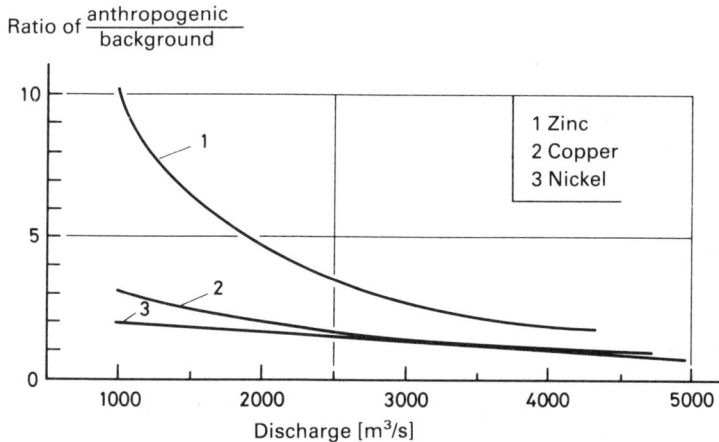

Fig. 7.3 — Changes in the relationship between heavy metals of anthropogenic and geogenic origin as a function of the discharge (schematic).

7.4 DISSOLVED AND UNDISSOLVED HEAVY METALS

As long as anthropogenic heavy metal inputs enter the receiving water at a constant mass flowrate and there contribute to a substantial increment in the background level, there will be a close concentration–discharge relationship. This limiting case is found in almost ideal form for chromium, under the conditions obtaining in the

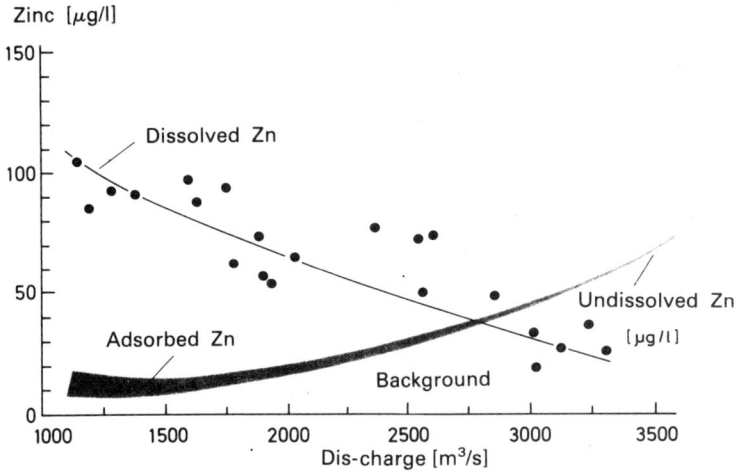

Fig. 7.4 — Concentrations of dissolved and undissolved Zn as a function of discharge in the
Rhine at Koblenz, 1972–74.

Middle Rhine in 1973/74 (Figs 7.5 and 7.6). The curves of similar pattern in the two
cases with approximately equal quotients for suspended solids/water of 250 000 are
indicative of a true equilibrium between dissolved and suspended solids. The
background value of 30 ppm inserted for the suspended solids in Fig. 7.6 is derived
from hydraulic calculations (see Part III).

Even more striking is the concentration–discharge relationship for zinc in 1979 in
the Lower Rhine (Fig. 7.7). There the above-mentioned effect (Section 7.3) is
superimposed, according to which the incipient flood flow (4310 m^3/s) is characterised by exceptionally high zinc and suspended solids concentrations. Leaving the
case of incipient flood flows out of account, an approximately hyperbolic decline in
the content of undissolved zinc is apparent in response to increasing values of the
discharge. The dissolved zinc concentration on the contrary appears to be practically
independent of the discharge.

The distribution of matter between the dissolved and undissolved fractions is not
merely affected by the discharge regime but also by the biochemical processes in the
water. The surprising decrease in the dissolved zinc and manganese between
February and July 1976 under dry weather flow conditions around and below the
MNQ value (Fig. 7.8) equals the decrease in the dissolved bicarbonate (Fig. 6.20)
and with a fair degree of certainty can be attributed to the same mechanism: the
phytoplankton production, linked with a shift in the Ca^{2+}/CO_2 equilibrium. In
agreement with this hypothesis, there is an increase in the Mn-content of the
suspended solids. The zinc content in the particulate fraction also showed a similar
increase. The co-precipitation of other metals, although not investigated in detail,
can likewise be inferred.

The favourable hydrological regime associated with the protracted period of dry
weather of several months' duration thus afforded an insight into the effects of the

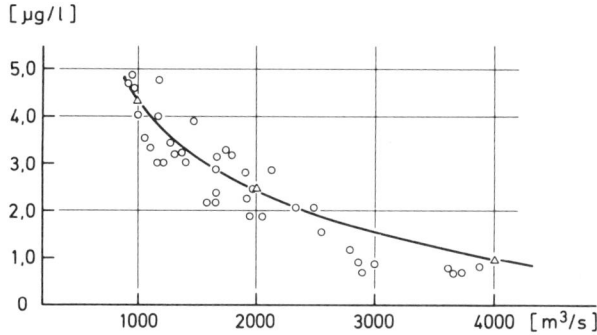

Fig. 7.5 — Dissolved Cr as a function of discharge in the Rhine at Koblenz, April 1973–December 1974 (from Schleichert, 1977/8).

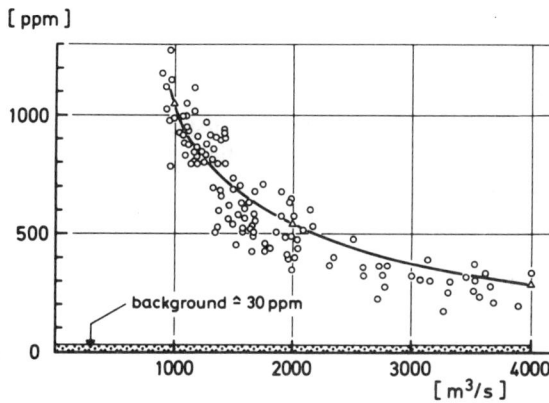

Fig. 7.6 — Undissolved Cr in suspended solids as a function of discharge in the Rhine at Koblenz, April 1973–December 1974 (from Schleichert, 1977/78).

biochemical processes on the trace contaminants. Even if the total concentration of a given metal in a litre of water was unaffected, nevertheless its distribution between the water and solids phases underwent a change. Depending on the particular objectives of trace metal analyses, the role of biochemical processes in the system should not be ignored.

7.5 DISCHARGE BEHAVIOUR AND LONGITUDINAL PROFILES

If one considers the heavy metal content of Rhine suspended solids at Koblenz in 1974 (Fig. 7.9) then there is for the insoluble chromium content a fairly close correlation with discharge, as already explained in Section 7.4. Not quite so closely

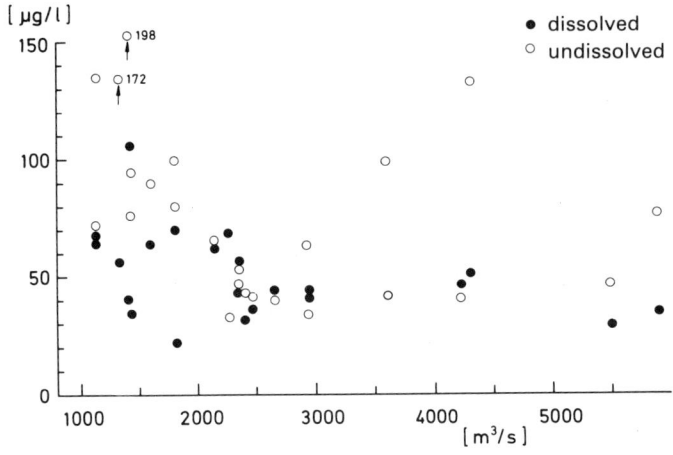

Fig. 7.7 — Dissolved and undissolved Zn as a function of discharge in the Rhine at Bimmen/Lobith, 1979.

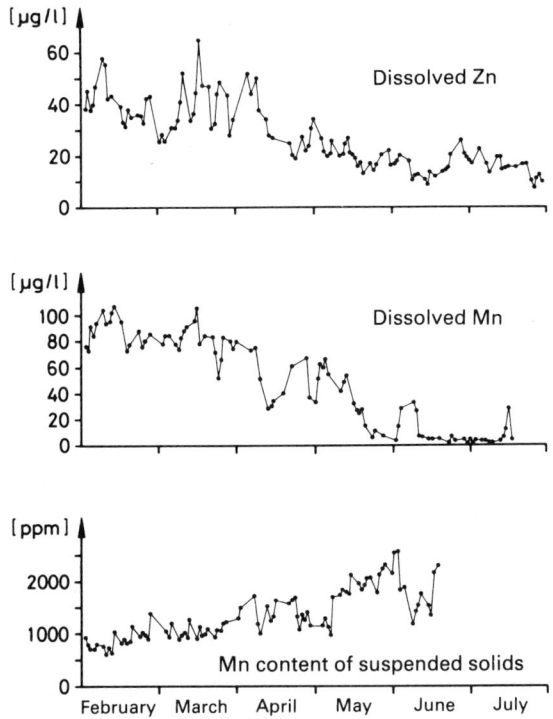

Fig. 7.8 — Dissolved Zn, dissolved Mn and Mn content of suspended solids in the Rhine at Koblenz, February–July 1976 (from Schleichert, 1977/78).

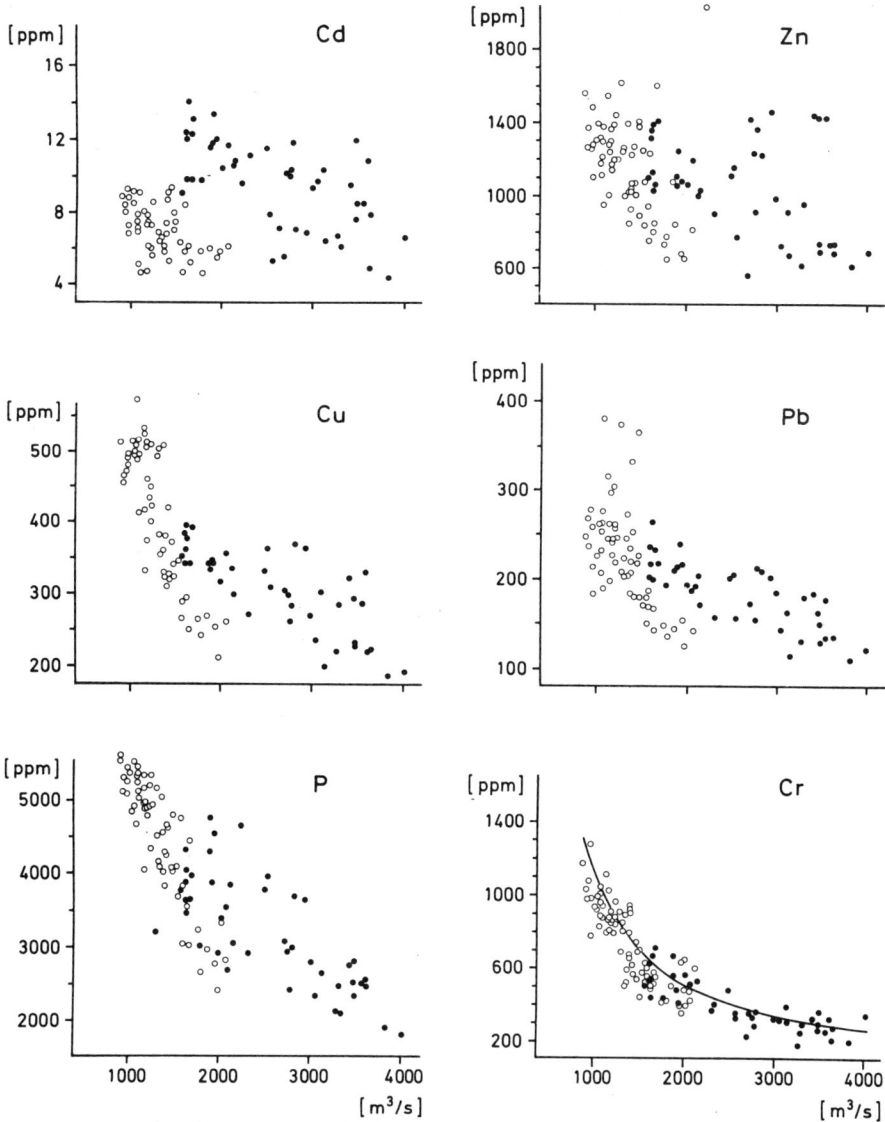

Fig. 7.9 — Correlations between discharge and the contents of heavy metals and phosphorus in the suspended solids fraction of the Rhine at Koblenz, May–December 1974 (from Schleichert, 1977/7). Key: ○ = May to 21 October, ● = 22 October to December.

correlated are the values for copper and lead, while cadmium and zinc appear to exhibit no relationship with the discharge at all. If one now takes into account the variation of discharge with time and origin then there is necessarily a distinction between the periods from May to 21 October and from 22 October to December. For the former the discharges were mostly around the MNQ mark, occasionally reaching the MQ level. They were derived principally from the Upper Rhine. Heavy rains led to an increase in the inputs from upland districts on and after 22 October, chiefly via the Neckar and Main, reaching flood proportions. They obviously carried into the Rhine large amounts of suspended solids of particularly high zinc and cadmium contents, somewhat elevated copper and lead contents, and chromium contents of almost exclusively geogenic origin. In addition the phosphorus contents were somewhat higher than the previous concentrations. We follow the details of Schleichert (1977/78) to the effect that, depending on the origin of the flows, two concentration–discharge relationships in succession were involved. The latter of the two encompassed the particularly hard-to-estimate inflow of highly polluted sediments from the regulated rivers. According to Fig. 2.6 they are formed at points of particularly low flow velocity and reduced entrainment action. Such a situation obtained during 1974 up to the month of October.

The analytical results correspond in the case under consideration with the particular hydrological conditions, and thus are quite intelligible. Moreover the results provide adequate confirmation of the reproducibility of the method of analysis, even if systematic errors cannot thereby be completely dismissed. Also, analytical results for a Rhine transect, in the downstream direction from Rhine to Emmerich–Lobith, provide evidence of the internal consistency of the measurements. For discharges of the MQ magnitude which were not noticeably influenced by rainfall events, the zinc concentration rose from 40 μg/l (dissolved) to 170 μg/l. The peak value for the total zinc content can be put down to the undissolved fraction, which from previous experience is the result of increased suspended solids inputs from tributaries such as the Neckar and Main. For Koblenz at kilometre 590.3 Fig. 7.10 indicates that 15 μg/l can be allotted to the undissolved fraction. If a suspended

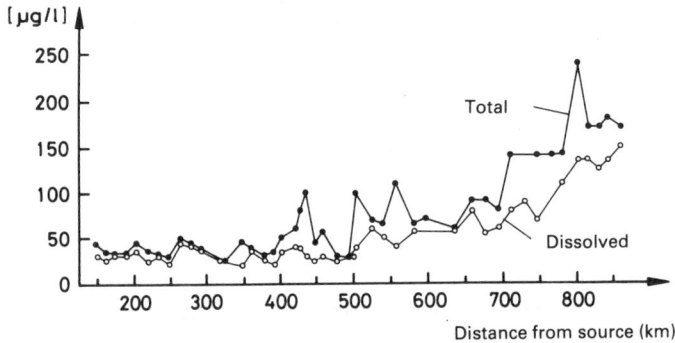

Fig. 7.10 — Dissolved and undissolved Zn along the length of the Rhine from Basle to Bimmen/Lobith, 24 June–1 July 1974 (International Commission, 1975).

solids concentration of 20 mg/l is assumed, the calculation indicates a suspended solids transport of 750 ppm, which together with the accompanying discharge of 1700 m^3/s is completely in accord with the results of Fig. 7.9 which are thus confirmed. As one can see, the dissolved zinc fraction outweighs the undissolved fraction over the entire stretch of the river, although both fractions are of roughly the same order of magnitude.

The analytical values for lead over the same stretch down to the 750 km mark are shown only as total lead, as the dissolved lead was below the limit of quantitative estimation of 10 μg/l, see Fig. 7.11. If one considers Koblenz again and assumes a

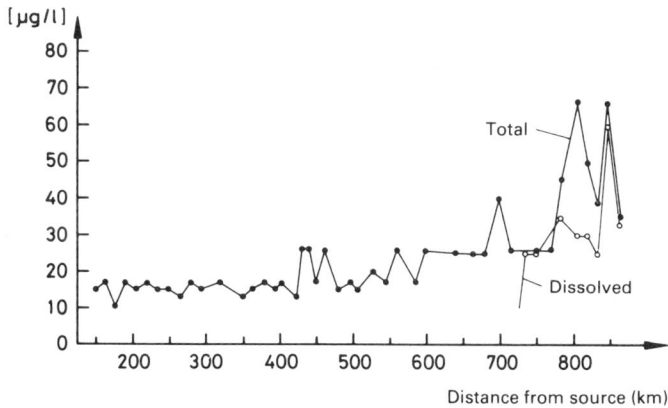

Fig. 7.11 — Dissolved and undissolved Pb along the length of the Rhine from Basle to Bimmen/ Lobith, 24 June–1 July 1974 (International Commission, 1975).

maximum content for soluble lead of 2 μgl, then 16 μg/l can be allotted to the undissolved fraction, corresponding to 800 ppm in the supended solids. This value does not agree with the value of *ca*. 150 ppm, which would be expected from an examination of Fig. 7.9 and also lies outside the range of normal lead concentration for Rhine suspended solids.

The longitudinal profile study of the Rhine from Constance to Lobith performed by another Task Force at the beginning of August 1977 gave soluble lead contents of 0.2 to 1.2 μg/l (Fig. 7.12). The improvements in analytical technique that had occurred since the earlier work were such that the values of around 5 μg/l reported previously would seem unlikely. The undissolved lead fraction is accordingly greater; in the downstream direction it increases, and more particularly in the Lower Rhine. The marked variations in the individual values (Fig. 7.13) are, if they are real, not understandable without reference to the mass transport of suspended solids. They moreover serve to highlight the problem of sampling. A concentration profile of this kind cannot be regarded as typical of the longitudinal profile of the Rhine.

From the examples presented, some probable generalisation can be derived. Severe fluctuations in the measured values, e.g. at a particular sampling point or in a longitudinal sequence, point either to defective analyses or to incorrect sampling methods. Smaller fluctuations within the same order of magnitude can, however, often be explained quite satisfactorily by reference to the hydrological conditions.

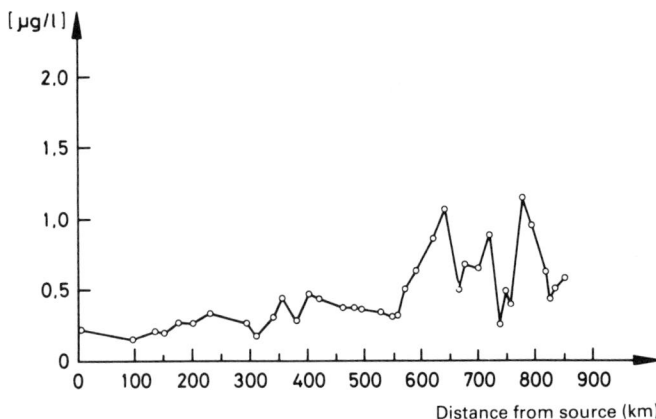

Fig. 7.12 — Dissolved Pb along the length of the Rhine from Constance to Bimmen/Lobith, August 1977 (from Breder, 1981).

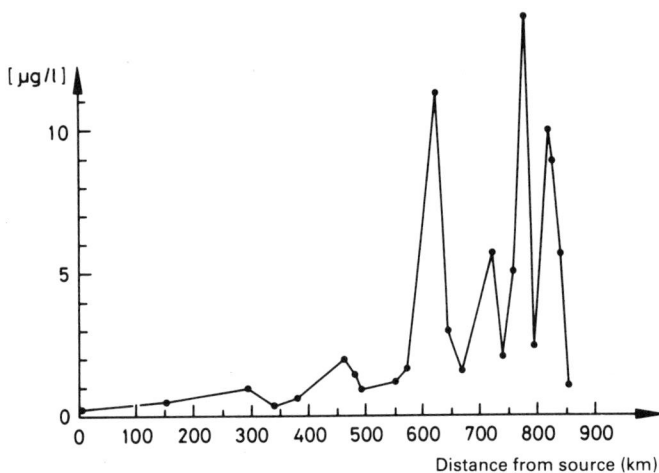

Fig. 7.13 — Undissolved Pb along the length of the Rhine from Constance to Bimmen/Lobith, August 1977 (from Breder, 1981).

7.6 SAMPLING OF SEDIMENTS FOR HEAVY METALS ANALYSIS

In the four-part sequence of Sampling–Sample pretreatment–Analysis–Interpretation each of the particular functions is important, but not necessarily of the same difficulty for every type of matrix. For sediments, if water flowrate, suspended solids contents and suchlike are disregarded, then an important reference framework for the analyst will be lacking.

Although sediments are derived from suspended solids, and likewise reflect the condition of the water body, they nevertheless differ, at least in flowing waters, in their 'specific composition'. In the suspended solids clay and silt fractions predominate (Fig. 2.4). Only during flood periods are sand particles intermingled with the other fractions, while in the sediments themselves the proportion of sand is usually greater. The pollution load of sediments is not necessarily governed by effluent discharges in the vicinity, as an exact relation between sources of pollutants and sediment pollution is rendered difficult or even impossible under flood conditions in flowing waters. Besides, the sedimentation conditions are often very changeable. According to Section 7.5, however, longitudinal profile studies can be very helpful in clarifying the origin of the trace metal contamination of suspended solids. For the sediments deposited more or less by chance at various points along the course of the river, to attempt such a relationship can be very questionable.

Some different problems of sediment analysis will now be examined with the help of three examples. Considering first of all inland lakes, where the incoming suspended solids are fractionated in the direction of travel, the finest solids are carried furthest into the lake and ultimately sink to the bottom. At the deepest places of Lake Constance (250 m depth) only the fine particulate fraction is present (Fig. 7.14). If one obtains a depth profile at this point one can determine the specific

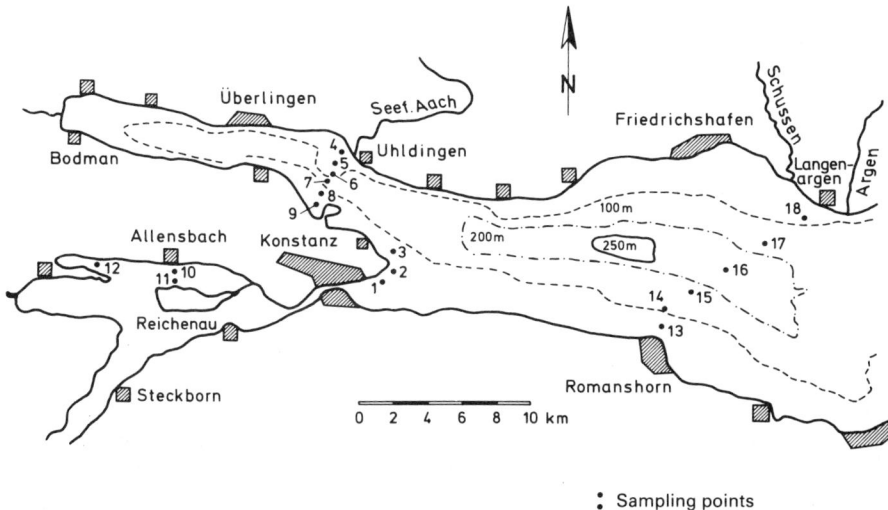

Fig. 7.14 — Water depth and sampling points in Lake Constance.

pollution of the entering solids and the pollution trends over a period of several decades, as Müller (1983) showed. The dating can be confirmed with reference to the content of artificial radionuclides e.g. Cs-137. For the numbered sampling points, about which more will be said in Chapter 8, such profiles are not always meaningful, especially where, among other things, sediment exchanges cannot be ruled out, and the sediment is not always uniform with respect to particle size. Thus in the entrance regions for the feeder streams (Nos 4 and 18) the proportion of sand is relatively large.

Basically a chronology of the pollutant inputs into the lakes appears a feasible task; the situation in Fig. 7.15 is, however, very much more complex. On the one

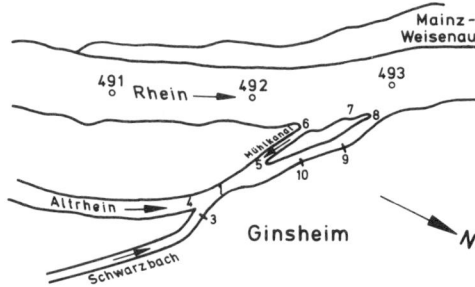

Fig. 7.15 — Problems affecting sediment studies in the old arm of the Rhine at Ginsheim.

hand for higher flow rates the suspended solids in the Rhine pass via the Mühl canal into the Old Rhine channel. This, however, also receives inputs of suspended solids at certain times from the upper reaches. The Schwarzbach, on the contrary, continually conveys highly polluted suspended solids, which then settle out in the entrance region to the Old Rhine (Nos 1 and 3 and to some extent also No. 10). Thus some exceptionally varied zones of pollution may be formed within a very narrow compass, the pollutants being derived from widely differing sources. The sedimentation conditions are governed by the flow of the Rhine into the Mühl canal and the Old Rhine. Only by means of numerous samples taken from various points is it possible to obtain a correct view of the trace metal contamination in the bottom sediments. In particular the suspended solids in the Schwarzbach may provide a case in point where the organic fraction and its metal contamination cannot be disregarded; a particle size fractionation is not advisable in this case.

It has already been shown in Figs. 2.7 and 2.8 that an apparently homogeneous sedimentation region is by no means homogeneous. In the following example the sediment deposits from a regulated reach were removed by a suction dredger and transferred to so-called drainage fields on dry land. This field on the right bank of the Lower Main is shown in plan in Fig. 7.16. On 16 October 1984, samples were taken from Areas II and III and analysed for heavy metals. According to Table 7.3 the concentration of all metals in the 300 m long drainage area increased from sampling point 1 to sampling point 38. For zinc the values rose from 87 to 1560 mg/kg, for lead from 16 to 201 mg/kg and for cadmium from 0.4 to 16 mg/kg. The head of the spreader was located in the vicinity of sampling point 1 (to the right of the figure). There as expected the sandy fraction of low trace metal content settled out. The clay fraction was transported far from the head of the spreader. These artificial hydraulic

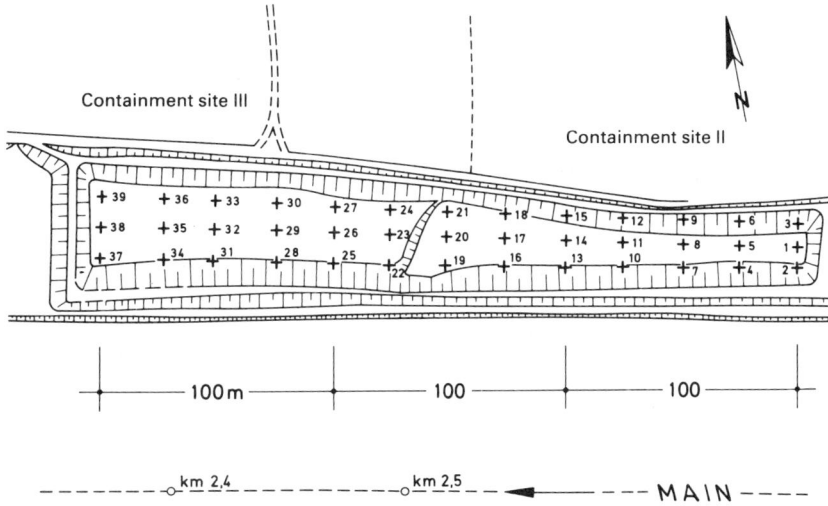

Fig. 7.16 — Fractionation of dredge spoil in a containment site on the Lower Main at Kostheim.

Table 7.3 — Heavy metal contents in soil samples from the Kostheim drainage fields
II and III taken on 16 October 1984 (results in terms of dry solids)

Sample	Heavy metal content (mg/kg)						
	Hg	Cd	Pb	Zn	Cu	Ni	Cr
1	0.47	0.4	16	87	22	11	28
5	0.43	0.2	17	94	16	12	52
8	0.30	0.2	22	59	14	10	8
11	0.66	0.2	18	81	20	15	21
13	0.86	1.0	64	277	56	40	56
17	1.2	1.4	50	279	53	32	93
20	1.6	3.0	86	486	87	59	90
23	3.2	5.6	109	724	126	70	102
26	2.2	3.2	119	728	127	66	90
29	2.3	3.5	114	742	134	70	90
32	2.4	4.5	118	782	153	75	89
35	6.6	19.1	213	1800	282	110	186
38	5.8	16.0	201	1560	255	104	188

conditions led to exactly the same result as those attributed to the flow regime in natural waters (Fig. 7.1).

How does one obtain representative samples under such conditions? The question gains in potency when the dredge spoil must be deposited on land for a long period and the soil limiting values of Kloke are to be applied. In the case of the drainage field referred to, the trace metal contents for samples 1–17 were below the limiting values, while those from No. 20 onward were above them. This material should therefore not be employed for either agricultural purposes or forestry. Likewise it is apparent what enormous, even financial, importance attaches to the sampling of sediments.

Apart from the practical examples referred to there frequently arises a problem of comparability of individual results as well as of data from different authors. As practically any sample possesses its own particle size distribution, one must standardise on a certain size fraction of < 0.06 or < 0.02 or < 0.002 mm. Dry sieving is not always successful and it is better to use a wet-sieving technique in an ultrasonic bath. According to Ackermann the fraction < 0.02 mm can be obtained using Perlon sieves, which should contain 90% of the total heavy metal load. Organic particles are thrown away. Over and above this a 'correction' of the trace metal content of the non-fractionated sample to the 'standard' < 0.02 mm fraction may be possible, certainly with the help of the caesium concentration (Ackermann 1983).

Overall, the breakdown of the analytical results into certain size fractions is to be welcomed. It is of no benefit if it is a matter of the Kloke limiting values, which apply to material of size < 2 mm although of a very wide particle size range. It is also not recommended in the case of organically polluted sludges, such as occur frequently in inland waters.

LITERATURE

7.1 and 7.2

Doerfel, K. (1962). Beurteilung von Analysenverfahren und -ergebnissen, Springer-Verlag, Berlin.

Förstner, U., Müller G. (1974). Schwermetalle in Flüssen und Seen, Springer-Verlag, Berlin Heidelberg New York.

Ackermann, F., Bergmann, H., Schleichert, U. (1979). On the Reliability of Trace Metal Analyses: Results of Intercomparison Analyses of a River Sediment and an Estuarine Sediment, Fresenius' Z. Anal. Chem. **296**, 270–276.

Hellmann, H., Griffatong, A. (1971). Beitrag zur optimalen Bestimmung im Wasser gelöster Schwermetalle durch Röntgenfluoreszenz, Z. Anal. Chem. **257**, 343–345.

Hellmann, H. (1973). Matrixeffekt und Korngrößenverteilung bei der Röntgenfluoreszenzanalyse von Feststoffen der Gewässer, Fresenius' Z. Anal. Chem. **263**, 14–19.

Bundesanstalt für Gewässerkunde (1983). Zusammenfassung der Ergebnisse der Ringanalyse "Bestimmung von Schwermetallen in Sedimenten", Koblenz, BfG-0162.

Hellmann, H. (1978). Selektive Bestimmung von Schwermetallen durch organische

Komplexbildner und Röntgenfluoreszenz, Fresenius' Z. Anal. Chem. **289**, 24–27.

Klärschlamm VO (AbfKlärV) vom 25. Juni 1982, in Kraft getreten am I. April 1983.

7.3

Rösler, H. J., Lange H. (1976). Geochemische Tabellen, Ferdinand Enke Verlag, Stuttgart, S. 278–283.

Hellmann, H. (1972). Definition und Bedeutung des backgrounds für umweltschutz-bezogene gewässerkundliche Untersuchungen, Dtsch. Gewässerkd. Mitt. **16**, 170–174.

Wedepohl, K. H. (1968). Origin and Distribution of the Elements, Pergamon Press, Oxford.

Winkler, H. A. (1978). Beurteilung der Schwermetallbelastung von Gewässern, Hydrochem. hydrogeol. Mitt. **3**, 203–222.

Kloke, A. (1977). Orientierungsdaten für tolerierbare Gesamtgehalte einiger Elemente in Kulturböden. Mitt. des Verbandes Deutscher landwirtschaftlicher Untersuchungs- und Forschungsanstalten H. 2, 32–38.

Baumann, A., Best, G., Kaufmann, R. (1977). Hohe Schwermetallgehalte in Hochflut-Sedimenten der Oker (Niedersachsen), Dtsch. Gewässerkd. Mitt. **21**, 113–117.

Fauth, H., Siewers U. (1983). Bäche–von der Natur selbst belastet, Bild der Wissenschaft **5**, 78–95.

Lichtfuß, R., Brümmer, G. (1981). Natürlicher Gehalt und anthropogene Anreicherung von Schwermetallen in den Sedimenten von Elbe, Eider, Tave und Schwentine, CATENA **8**, 251–264.

7.4 and 7.5

Schleichert, U. (1977/78). Bilanzierung und Herkunft toxischer Schwermetalle für das Rheineinzugsgebiet, Berichte über Landwirtschaft **55**, 691–699.

Internationale Kommission zum Schutze des Rheins gegen Verunreinigung (1975). Langfristiges Arbeitsprogramm (LZP)—Gesamtfassung, Koblenz.

Breder, R. (1981). Die Belastung des Rheins mit toxischen Metallen, Dissertation der Rheinischen Friedrich-Wilhelms-Universität Bonn.

Arnold, H., Gundlach, H., Schweisfurth, R. (1982). Untersuchungen zur Anreicherung von einigen Spurenelementen in Manganoxiden, Vom Wasser **58**, 165–172.

7.6

Müller, G. (1983). Zur Chronologie des Schadstoffeintrags in Gewässer. Geowissenschaften in unserer Zeit **1**, 2–11.

Bundesanstalt für Gewässerkunde (1970). Die Ermittlung der Herkunft der Schlammabalgerungen im Ginsheimer Altrhein, Gutachten zu G1/255-Ginsheim 8358, Koblenz.

Ackermann, F. (1980). A Procedure for Correcting the Grain Size in Heavy Metal Analyses of Estuarine and Coastal Sediments, Environ. Techn. Letters **1**, 518–527.

Ackermann, F. (1983). Einführung in die Problematik der Korngrößenkorrektur im

Zusammenhang mit Schwermetallanalysen in Sedimenten, Interner Bericht, BfG, Koblenz.

Züllig, H. (1982). Die Entwicklung von St. Moritz zum Kurort im Spiegel der Sedimente des St. Moritzer Sees, Wasser, Energie, Luft **74**, 177–183.

Knöpp, H. (1984). Baggergutprobleme an den Bundeswasserstraßen. Fachseminar Baggergut, Hamburg.

Müller, G. (1979). Schwermetalle in den Sedimenten des Rheins—Veränderungen seit 1971. Umschau **79**, 778–783.

Deutsche Forschungsgemeinschaft (1982). Schadstoffe im Wasser, Bd. I Metalle, Forschungsbericht.

Länderarbeitsgemeinschaft Abfall (LAGA) (1980). Richtlinien für das Vorgehen bei physikalischen und chemischen Untersuchungen im Zusammenhang mit der Beseitigung von Abfällen, LAGA-AAM 21/80; SM 2/79—Bestimmung von Schwermetallen in festen und schlammigen Abfällen.

Siehe auch Jahresbericht der ARW, insbesondere Heft 1979.

8

Hydrocarbons

8.1 PRELIMINARY REMARKS

The determination of hydrocarbons has, since the beginning of the 1960s, constituted one of the regular tasks of surface water analysis. The outward occasion for this was the increasing consumption of petroleum products. For the analytical 'determination of oils and fats' initially the procedures of DEV Vol. 17–18 were adopted. Petrol ether was employed as the extractant. With the arrival of IR spectroscopy this was superseded by tetrachloromethane (CCl_4) and later by 1,1,2-trichloro-trifluoroethane.

Owing to three decisive pitfalls, the already difficult task of hydrocarbon analysis was rendered much more exacting. The first involved the use of incorrect terms. Many analysts reported the materials contained in the CCl_4-extract as 'total hydrocarbons', and not correctly as 'total extract'. As CCl_4 had a more pronounced action on the polar constituents than the petrol-ether formerly employed, frequently larger quantities of vegetable and animal fats, emulsifiers and waxes were extracted. After separation of the polar substances by adsorption on Al_2O_3, one consistently termed this part of the extract 'polar hydrocarbons', with which the non-polar 'paraffin hydrocarbons' were contrasted. As certain expert opinions of oil spills demonstrate, this confusion of terms still has unfortunate effects today.

The second pitfall, for which there may be some excuse, concerned equating hydrocarbons with mineral oils. Only very slowly did the existence of hydrocarbons of recent biogenic origin begin to make its mark in the literature. Concerning the differentiation between the two fractions, more detailed consideration is needed. That water analysts could claim, then and now, only slight knowledge of the composition of different petroleum products may likewise be worthy of comment.

The third pitfall arose because people were ignorant, or insufficiently aware, of gradual changes in the composition of different mineral oils, supplied in the form of 'specimens' of particular compounds, as well as of the different behaviour of alkanes and aromatic hydrocarbons in natural waters. Thus when interpreting the results, one always referred back to the composition of the original oil. Although the subject of oil analysis has been widely discussed in the literature, there are still no systematic studies of this nature.

8.2   BRIEF DETAILS OF ANALYSIS

Even if one disregards the special subject of oil identification in the context of polluter detection (Hellmann 1973), there is still an enormous volume of literature on the subject of oil in water. Accordingly it will be enough if at this point — as in the case of heavy metals — the method of analysis is referred to only briefly.

By far the most widely used method of analysis of hydrocarbons in surface waters is group determination by means of IR spectroscopy (DIN 38 409, Part 18). The interfering contaminants are removed by adsorptive methods. In connection with the further fractionation of hydrocarbons on suitable adsorbants, such as silica gel, aluminium oxide or Sephadex LH-20, alkanes and aromatic hydrocarbons may be separately determined. They can then be quantitatively assessed on the basis of IR spectroscopic results.

The adding process does not always give absolutely correct results. Problems exist in connection with the treatment of aromatic rings and CH-groups, as well as quaternary C atoms. These must also be incorporated in the calibration factor. The result is more reliable the closer the reference substance is in its composition to that of the hydrocarbon mixture under test. It is not always possible, on the basis of the spectrometrically determined alkane content (extinctions at 2930 and 2960 cm^{-1}), to incorporate the aromatic fraction in the calibration factor. On the other hand the aromatics can be determined with a high degree of sensitivity by means of their fluorescence in hexane as solvent. For this purpose the wavelength combination of 310–360 nm is employed. The result applies specifically to the aromatics 'by themselves' and one cannot include the alkanes in this operation. In practice the method is complicated by the presence of numerous components of similar fluorescent properties such as the polycyclic aromatic hydrocarbons.

A further group determination can be effected by thin-layer methods, frequently on silica gel F-60. The alkanes are rendered visible with the aid of spray reagents (e.g. Merck) and evaluated with reference to the spot size. Aromatics can also be determined by fluorescence detection. If one overlooks for a moment the fact that the results may be considered as only semi-quantitative, this method is capable of supplying very valuable information concerning the composition of the extract; examples are given in Section 8.3.

Besides IR spectroscopy the method of gas chromatography is very widely employed. After the previously mentioned preliminary separation on absorbent materials, the alkanes and aromatic fractions in the low and medium boiling ranges may be largely separated into the individual constituents. Thus one obtains information about the composition of these fractions, and possibly the mineral oil types and boiling ranges. Using special detectors in addition to the customary flame ionisation detector, fractions containing hetero-atoms such as S and N may be identified, which may also contribute to the characterisation of the sample.

From the instrumental viewpoint it is also possible to obtain fluorescence spectra (synchronous, excitation and emission spectra) for sample extracts. Their use is, however, limited in the context of surface water analysis.

Not very informative, even if sometimes very useful, are IR-scanning spectra in the range from 60 to 4000 cm^{-1}.

Finally a remark may be made on the specific circumstances obtaining when taking the sample. Oil as a separate phase, that is oil floating on the surface, can never be so accurately removed as to ensure a correct value for the amount. Certainly oil separating devices operating on the principle of an oil separator are preferable to adsorbent materials (absorbent cubes). However, the analyst is chiefly concerned with the identification of the oil rather than with its exact quantity. The dissolved oil fraction is extracted after removal of the suspended solids. The suspended solids must also be extracted as for sludges and sediments; for more details see Sections 8.4–8.6. The separate determination of hydrocarbons in water and sediments is particularly indispensable, where the suspended solids content is high (>50 mg/l).

8.3 HYDROCARBONS OF BIOGENIC AND MINERAL ORIGIN

Leaves and flowers of terrestrial plants invariably contain alkanes as well as certain unsaturated hydrocarbons, principally in the form of straight-chain paraffins in the boiling point range between n-C_{20} and n-C_{33}. A very characteristic feature is the predominance of certain members of uneven C-number, such as the compounds C_{29}, C_{31} and C_{33} (Fig. 8.1) or C_{23} and C_{25} (Fig. 8.2). In addition, photosynthetic

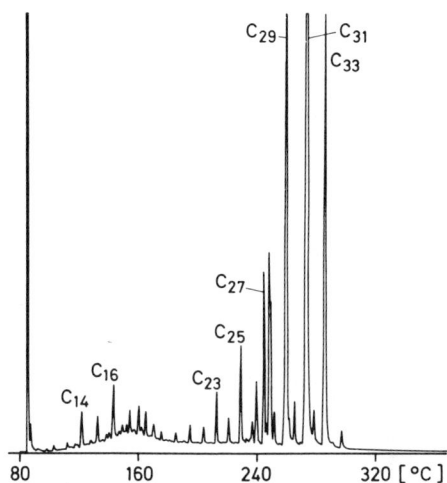

Fig. 8.1 — Gas chromatogram of the alkane fraction of grass. Packed column, 3 m, K/NaNO₃ eutectic.

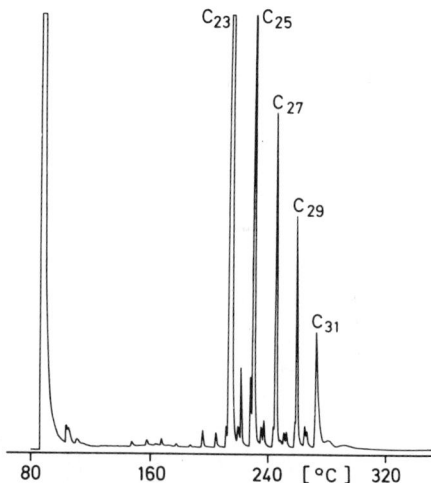

Fig. 8.2 — Gas chromatogram of the alkane fraction of forsythia flowers.

organisms in natural waters such as phytoplankton produce alkanes and alkenes, although with a different C-distribution, in which n-C_{17} compounds predominate. The amount of hydrocarbons in terrestrial plants can reach about 1⁰/₀₀ of the dry weight, and in algae up to 5⁰/₀₀. On the basis of a rough scale-up, this implies an annual production worldwide in the first case of 800 million tonnes, or about 40% of the entire world oil supply.

The typical distribution pattern of biogenic hydrocarbons, by which one under-

stands the occurrence of the respective compounds and their proportion of the total
hydrocarbon content, is recognisably different from that of the mineral oil products
(for this see Section 8.12). But not only the small proportion of *iso*-alkanes, but even
more the absence of typical mineral oil aromatics, is the characteristic feature of
biogenic hydrocarbons. This can be very quickly recognised following a chromatographic separation on silica gel plates. After development in a trough saturated with
n-hexane the plates are sprayed with Rhodamine B (0.03% in H_2O) and observed
under ultraviolet light (300–400 nm) (Figs 8.3 and 8.4). If one sprays the developed

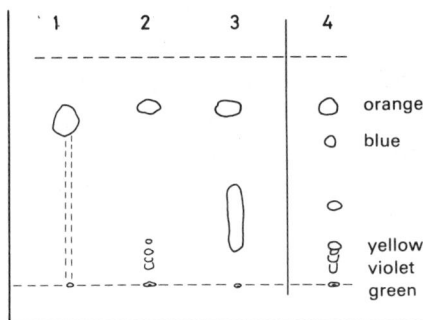

Fig. 8.3 — Characterisation of different hydrocarbon mixtures on silica gel F-60 — No. 1
Rainwater extract; No. 2 Extract of plants; No. 3 Heating oil El; No. 4 as for No. 2 but sprayed
with Rhodamine and visualised in long-wave UV. Solvent: *n*-hexane.

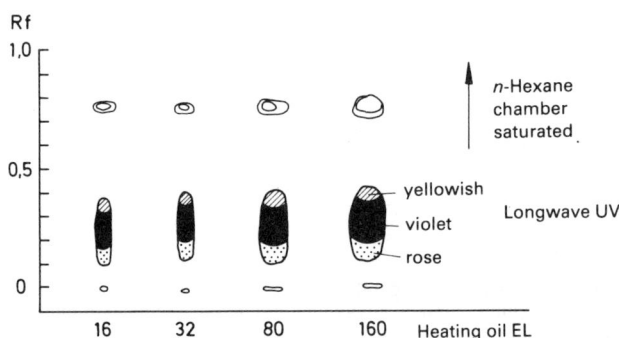

Fig. 8.4 — Fractionation of heating oil El on silica gel F-60.

thin layer not with Rhodamine B but with diluted permanganate solution, then only
the aromatics, and under certain conditions individual unsaturated compounds,
become visible as yellow spots on a violet background.

Where unsaturated biogenic compounds are isolated in addition to alkanes, one
can also characterise these with the aid of gas chromatography. Mineral-oil aroma-

tics give rise to a different type of chromatogram, which exhibits a homologous series of alkylbenzenes, napthalenes, phenanthrenes/anthracenes, etc., with a symmetric frequency distribution.

Finally, mineral oil aromatics may also be positively identified with the aid of IR spectroscopy. For this purpose one chromotographs the extract ideally in two stages, each in unsaturated enclosures, as described in Fig. 8.5. The aromatic fraction with

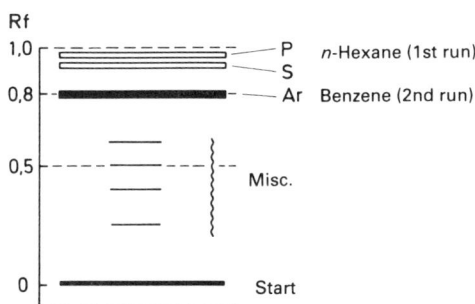

Fig. 8.5 — Fractional chromatography of a sample extract on silica gel in an unsaturated chamber. Key: P=paraffins, S=sulphur, Ar=aromatics.

an R_f value of *ca.* 0.8 is separated from the support, along with the silica gel, and eluted with chloroform or cyclohexane.

With regard to the pseudo-aromatics (i.e. unsaturated hydrocarbons) in the biogenic extracts, one proceeds similarly.

8.4 HYDROCARBONS IN NATURAL WATERS — WATER SAMPLES

In general, it is not to be expected that the biogenic nature of hydrocarbons will remain unaffected during entry into soil and water, because at least from time to time petroleum products will also become involved. For the analyst, therefore, the situation sketched in Fig. 8.6 prevails. The group determination of hydrocarbons (the measured reading in Fig. 8.6) *by itself* gives no clue as to the origin of the material. Even though one is aware from experience that elevated values, e.g. over 0.5 mg/l, must be partly composed of petroleum hydrocarbons, and the more so as the level increases, yet on the other hand it is by no means an easy task to determine the background level for hydrocarbons of biogenic origin in natural waters, at least with any degree of certainty.

Of the many possible approaches to the illumination of this 'grey zone' which perhaps ought to be called the 'blackground', one possibility has already been mentioned, namely that of checking for the presence of mineral oil aromatics in addition to alkanes. A negative result for rainwater was indicated in Fig. 8.3 so that one may regard the alkanes in this case — subject to any further indications to the contrary — as being of purely biogenic origin.

The above chain of thought carries the implicit assumption that mineral oil

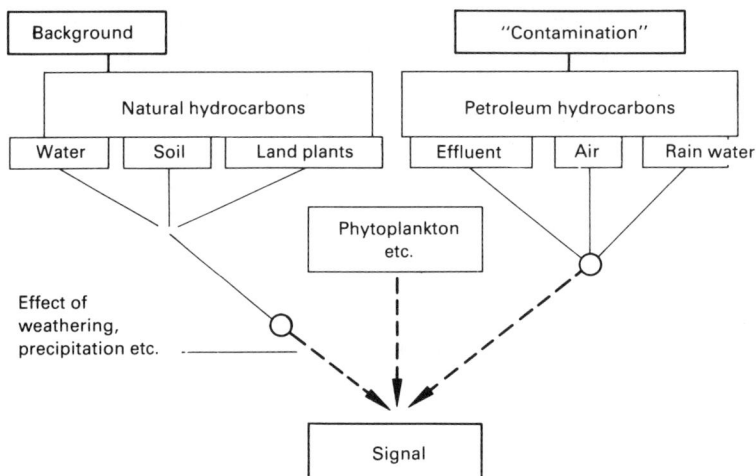

Fig. 8.6 — The make-up of the instrument signal in the course of hydrocarbon group analyses in natural waters.

aromatics are not subject to much faster photochemical or biochemical decomposition processes in water and sediments than alkanes. For complete clarification of this topic further investigations would be desirable.

A frequently used method endeavours to assign the alkanes according to their origin on the basis of the gas chromatogram. In the chromatogram of the alkane fraction of spring water reproduced in Fig. 8.7, the *n*-paraffins are numbered. The

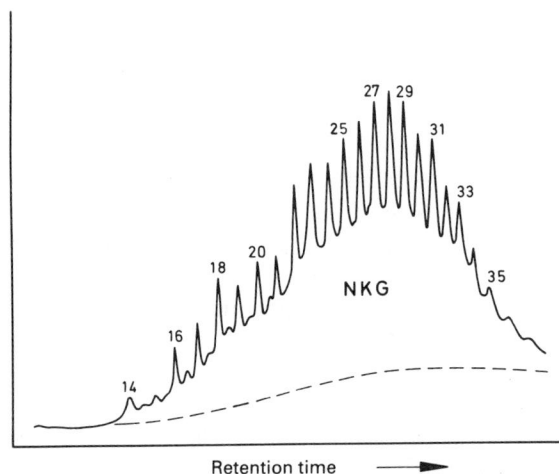

Fig. 8.7 — Gas chromatogram of the alkane fraction in spring water (Sieg spring, 18 January 1973). Packed column, silicone rubber.

boiling-point range extends from n-C_{14} to beyond n-C_{35} with a maximum at n-C_{28}. The compounds are practically symmetrically distributed about the maximum *without alternating*, i.e. without any preferential occurrence of odd numbered C values. This and the unresolved background (NKG=non-separable complex mixture) is, according to many authors, a characteristic feature of mineral oils. It thus gives rise to a contradiction in the interpretation of the results, since mineral oil aromatics could not be detected.

Regarding the question of the 'non-separable background', the experts disagree. What is not in dispute, however, is that the biochemical decomposition of biogenic hydrocarbons in natural waters will lead to the formation of just such a background. Thus the addition of a hydrocarbon extract of purely vegetable origin of the kind shown in Figs 8.1 and 8.2, to a model system containing Rhine water, led after a few days to an extract composition of the type indicated by the chromatogram in Fig. 8.8.

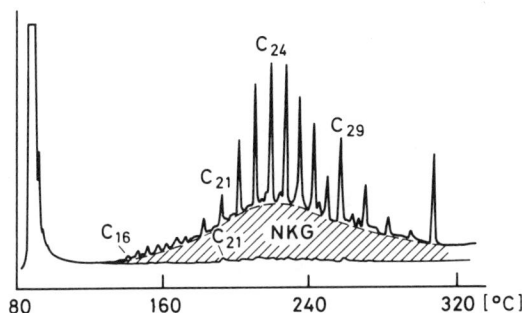

Fig. 8.8 — Gas chromatogram following partial diagradation of biogenic hydrocarbons in Rhine water. Packed column as in Fig. 8.1. Lower line=blank value of water.

Also with the superior resolution of a capillary column compared to that obtaining for Fig. 8.7, an unresolved background in the high-boiling region is revealed. Chromatograms with non-separable background fractions are invariably obtained from an analysis of the alkane fraction of sediments. As a result any demonstration of the origin of the hydrocarbons by means of gas chromotography is still accompanied by question marks.

In the past efforts were made to characterise the alkanes according to their origin from a measurement of the specific ^{14}C activities (Kölle/Stieglitz, 1974). This method is to our knowledge still unconfirmed. Moreover it is outside the normal range of capability for routine analyses in a modern laboratory. From many years' experience, the concentration of biogenic hydrocarbons in spring waters, pristine streams and lightly polluted surface waters with relatively low suspended solids transport, is usually below 0.1 mg/l, and often less than 0.05 mg/l. Values several times greater than this are grounds for suspecting the presence of mineral oils.

More heavily polluted stretches and those severely polluted by effluent discharges where the suspended solids content must exceed 20 mg/l, as well as those exhibiting a markedly increased suspended solids load after rainfall events, may be distinguished by hydrocarbon contents greater than 0.1 mg/l, without there being any direct evidence of the presence of mineral oils. The stated values are to be understood as the sum of the dissolved and suspended solids-bound fractions. As hydrocarbons can be biochemically decomposed, both the type of concentration profile referred to in connection with heavy metals and also effects due to discharge and water temperature are to be expected.

According to Fig. 8.9 a decrease of the hydrocarbon content is clearly discernible

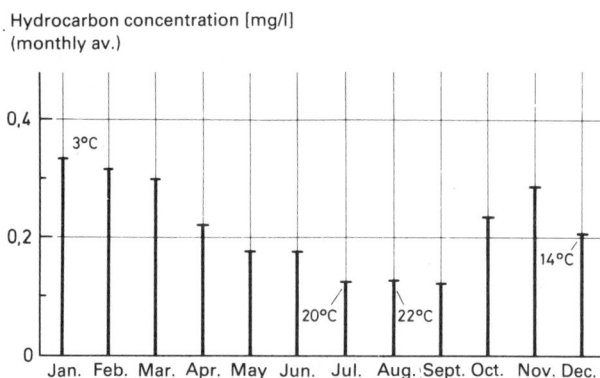

Fig. 8.9 — Effect of water temperature on hydrocarbon concentration in the Rhine at Koblenz, 1972.

from an examination of monthly averages. By means of chromatography it can be demonstrated that the straight-chain alkanes are preferentially degraded (Hellman, *Tenside*, 1973).

For hydrological investigations of the kind described in Part I of this book one should separate the suspended solids from the water phase and analyse them separately. Suspended solids during periods of high flow are usually more lightly polluted than those during low flow periods. For the Rhine at Koblenz in 1984, specific suspended solids contamination levels ranged from 1000 to 3500 mg/kg dry wt. The specific contamination increased with the proportion of organic matter, indicated by the loss on ignition for the various samples. In purely general terms the loss on ignition may serve — as in the case of heavy metals — as a tracer parameter for the magnitude of the specific pollution load.

If one relates the hydrocarbon content to the organic fraction of the suspended solids, then values such as those for the Rhine at Koblenz in 1983–84, from 8–18°/$_{oo}$ are obtained. Possibilities for interpretation are discussed in the following sections.

8.5 HYDROCARBONS IN LIGHTLY POLLUTED SEDIMENTS

In the surface waters of the German Federal Republic, it is always possible to detect the presence of alkanes, allowing for a possible enrichment during the course of the analyses. As a rule the typical mineral oil aromatics are absent, although with the aid of sensitive analytical methods some polycyclic aromatic hydrocarbons (PAHs) may be detected. Among these, fluoranthene is present in by far the largest amount. The concentration of PAHs as a group, however, lies about three orders of magnitude below that of the hydrocarbons.

If one assumes that the biogenic hydrocarbons of land plants, in so far as they are not partially volatilised, fall to the ground in the leaves, then one would expect to find them in sediments as a consequence of erosion processes and atmospheric inputs. Similarly, this also applies to the biogenic hydrocarbons of the phytoplankton. In those sediments which are not subject to massive pollution from effluents (Section 8.6) the hydrocarbon fraction should definitely be mainly of biogenic origin. In recent sediments from the Bienhorn brook at Koblenz, obtained in 1975, the alkane fraction complied fully with these expectations, according to Fig. 8.10. Boiling

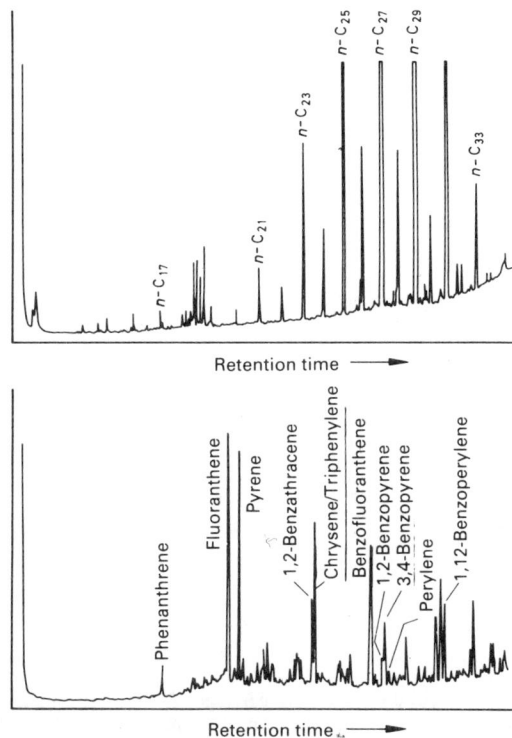

Fig. 8.10 — Sediments of recent (unpolluted) origin. Gas chromatogram of the alkane fraction (top) and the aromatic fraction (bottom). Capillary column, OV101 (from DGMK, 1977).

points, alternation of individual compounds with a preponderance of odd-numbered C compounds, and concentration maxima for C_{27} and C_{29} were confirmation of the biogenic character. In the aromatic fraction, mineral-oil aromatics were accordingly absent while in the PAHs from phenanthrene to beyond 1,12-benzo-perylene those with *unsubstituted* benzene nuclei predominated.

The most strongly occurring compound — as in the case of water samples — is fluoranthene. The quantities of alkanes and PAHs are in a proportion of 50:1. The hydrocarbon content of many similar sediments, expressed in terms of their content of organic matter (i.e. loss of ignition) is around 3–5°/$_{oo}$. It thus lies below the level of organic contamination of Rhine suspended solids at Koblenz in 1983–84 by a factor of 3.

The gas chromatograms of such samples are quite singular and easy to interpret. The other extreme, in which solids of high mineral-oil content are present, is analytically well demonstrated with reference to street sweepings. In Fig. 8.11, for

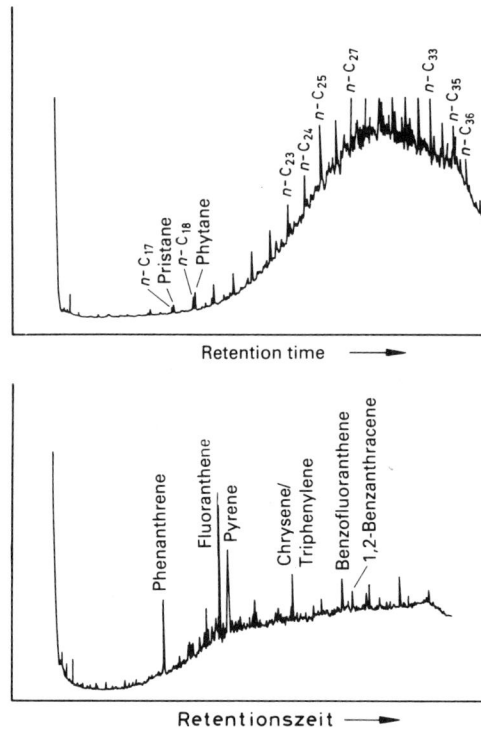

Fig. 8.11 — Street dirt from the edge of the carriageway. Gas chromatogram of the alkane fraction (top) and aromatic fraction (bottom) (from DMGK, 1977).

both the alkane and the PAH chromatograms, there is a large non-separable background fraction. A further distinctive feature is the extension of the boiling point to far beyond the n-C$_{33}$ compound. Even capillary columns are unable to achieve further resolution of this highly complex mixture. In other respects the mineral-oil aromatics can be detected more quickly and also more conclusively with the aid of thin-layer chromatography (see Chapter 9).

The sediments of Lake Constance can be taken as typical of sediments of a low level of anthropogenic pollution, as long as one disregards those aspects connected with the process of eutrophication and also adopts the situation prevailing in highly polluted flowing waters as a benchmark (Section 8.6). The positions of the sampling points (Fig. 7.14) are scattered over a wide area of the lake. The results are also very divergent — see Table 8.1. The material obtained from depths of 0–50 cm beneath the sediment surface is not very uniform with respect to either its particle size or its chemical composition. Fine particulate material with a relatively high loss on ignition (shown as organic matter in col. 4 of Table 8.1) is mostly also relatively rich in hydrocarbons. Nevertheless the highest value at 207 mg/kg lies far below the figures customary for suspended solids in natural waters. The specific sediment pollution is also correspondingly low, with a maximum of $3.2^o/_{oo}$ determined on the organic fraction.

In the middle of the lake (Table 8.2) a number of sediment cores were obtained from the bottom in a depth of 250 m of water and subjected to analysis. A depth profile for the year 1974 (Table 8.2) shows the increase in the hydrocarbon content between the deeper, older sediments and the most recent deposits. The heavy metals and also the phosphorus content show similar trends, increasing from earlier to recent times. According to very intensive studies by a group of analysts (DMGK 1982) hydrocarbon contents of 78 and 106 mg/kg were observed at the sediment surface in this, the deepest part of Lake Constance. Although these values, like those of the author, lie far below the relevant values for flowing waters, the question of mineral-oil contamination is nevertheless a subject of intense discussion.

The gas chromatogram of the alkane fraction (Fig. 8.12 top) shows once again an ominous non-separable fraction which extends from the C_{20} to the C_{30} paraffin region (cf. Figs 8.6 and 8.7). If even a partial decomposition of hydrocarbons in sediments takes place continually, then the C-value distribution in principle remains unchanged, so that the middle distillate can still be distinguished from lubricating oils after many years (Michael *et al.*, 1975). The corresponding chromatograms of a diesel fuel oil and a two-stroke mixture and their relevant non-separable fractions can also be seen in Fig. 8.12. The shape of the chromatogram for the lubricating oil strongly resembles that for the alkane fraction of the street sweepings in Fig. 8.11. Without wanting to complicate the matter further at this point, it may also be remarked that the results of analysis of the aromatic fraction indicate that chiefly four- and five-ringed aromatics in unsubstituted condition were present; for their occurrence pyrolytic processes were assumed to be responsible.

These findings do not at all coincide with, e.g., Meyer's reference book *The Human Environment* (1975) which on p. 354 states 'Nowadays every square metre of the lake floor is contaminated on average with 1 g of oil. Overall it is estimated that roughly 600 t of mineral oil have been deposited on the bottom of Lake Constance.'

8.6 HYDROCARBONS IN SLUDGES

As in the case of heavy metals, so with hydrocarbons, the question of sampling presents a problem. Closely linked with this is the question of particle sizes, i.e. the close relationship between the specific hydrocarbon content of a sample and the particle size distribution. A differentiation between the fractions of biogenic and anthropogenic (mineral-oil) origin is also required in this case. Over and above this, the hydrocarbon analysis appears to present even greater demands. The concentration in the sample may frequently be so high (for the solids it will be in the ppm

Table 8.1 — Analytical results for Lake Constance sediments, 1973

Location of sample	No.	Moisture content (% dry wt)	Organic matter (%)	CaCO$_3$ (% of loss on ignition)	Hydrocarbons (mg/kg wet wt)	Hydrocarbons (mg/kg dry wt)	(‰ org matter)
Peninsula	1	25	1.5	65.0	11	14.6	1.0
east of	2	36	2.1	57.4	11	17.2	0.8
Constance	3	39	2.7	72.8	5	6.0	0.3
Transverse	4	39	5.2	16.0	100	164	3.2
profile	5	48	5.0	22.4	66	126	2.5
	6	69	8.7	23.5	82	201	2.3
	7	71	8.1	28.5	33	114	1.4
	8	68	7.0	20.0	33	105	1.5
	9	65	5.9	25.0	16	45	0.8
Gnadensee	10	47	2.0	81.2	11	20	0.7
	11	68	6.0	70.0	11	51	0.8
	12	50	3.2	55.0	33	66	2.1
Transverse	13	62	4.6	40.6	11	29	0.6
profile	14	55	4.5	28.0	8	18	0.4
Romanshorn–	15	53	6.3	35.9	11	23	0.4
Langenargen	16	71	6.7	36.4	60	201	3.1
	17	63	5.8	38.9	60	162	2.8
	18	62	6.8	25.9	71	187	2.7

Moisture content 105°C for 48 h.
Organic matter 550°C for 1 h.

Table 8.2 — Analytical results for a sediment core from the centre of Lake Constance, 1974

Sediment depth (mm)(%)	Moisture content (% dry wt)	Loss on ignition (% dry wt)	CaCO$_3$ (% dry wt)	Phosphorus and heavy metals				Hydrocarbons (mg/kg)
				P	Zn	Pb	Cu	
				(mg/kg)			(mg/kg)	
0–3	80	9.5	40.0	822	235	117	23	164
20–40	64	8.1	34.5	539	123	55	23	58
50–70	59	6.3	33.2	506	105	54	22	55
70	59	5.9	29.2	448	83	45	21	38

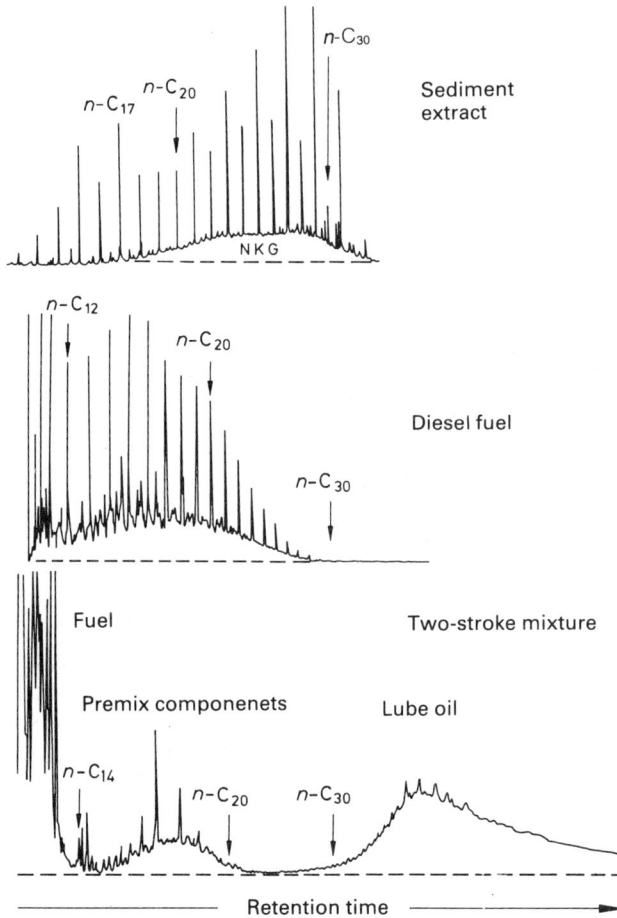

Fig. 8.12 — Alkane fractions of a sediment extract from Lake Constance — mid-point 1981 —
and of diesel fuel and two-stroke mixture (from Michael *et al.*, 1975, Völtz).

region) that the total content, within the limits of accuracy of a group analysis, can be determined with some certainty. However, the question of the background becomes more difficult to settle. In the case of recent sediments, the natural content of hydrocarbons is around $3^o/_{oo}$ of the loss on ignition, sometimes more, sometimes less. It is, however, not constant but diminishes in the older aerobic sediment layers owing to biochemical degradation — see results for Lake Constance sediments in Table 8.1. The situation is different for heavily organically polluted sediments or sludges in an anaerobic environment. There one may find concentrations of hydrocarbons in the organic matter which can exceed 10%. If one reckons with an organic fraction of 20% in the sample (based on dry wt), this indicates a specific contamination in the relevant samples of a total of 2000 mg/kg. Anyone who analyses highly contaminated sediments and suspended solids will have to cope with hydrocarbon contents of this order of magnitude.

According to the results of gas chromatography, the typical biogenic pattern of the alkanes of land plants or phytoplankton is only weakly exhibited, if at all. Equally, however, one should not attribute these high contents to mineral-oil contamination without further examination. In the field of geosciences it has long been known, for example, that fungi and bacteria transform organic matter in the sludge, with the formation of hydrocarbons (Krejci–Graf, 1959). As an intermediate stage in this conversion process 'bitumen' is formed. In the plankton of the Black Sea and the Caspian Sea hydrocarbon contents of $4–7°/_{oo}$, sometimes as high as $62°/_{oo}$, were recorded, which could likewise be included in the 'bitumen'.

Unfortunately studies of the bacterial production of hydrocarbons in sludges are lacking. That besides the recent biogenic background there is also a geogenic fossil background, which for argillaceous rocks may amount to 50 mg/kg, is of no account for the evaluation of hydrocarbon contents of sludges but does belong to the wider argument. Of considerable importance, however, are the hydrocarbons produced by microorganisms from other compounds, which may be barely distinguishable from mineral-oils.

On the evidence of the analyses, it seems that microorganisms do not produce any hydrocarbons analogous to the group of mineral-oil aromatics. Besides the absence of alkanes there are decreasing amounts of alkylbenzenes, napththalenes, phenanthrenes/anthracenes and higher alkylated aromatic nuclei in homologous series. The invariably occurring unsubstituted PAHs are attributed in the technical literature to the combustion of fossil fuels, chiefly coal and oil (i.e. PAHs of pyrogenic origin).

To what degree the hydrocarbons are present in the various solids contained in natural waters can be inferred from Fig. 8.13. For the analytical determination one

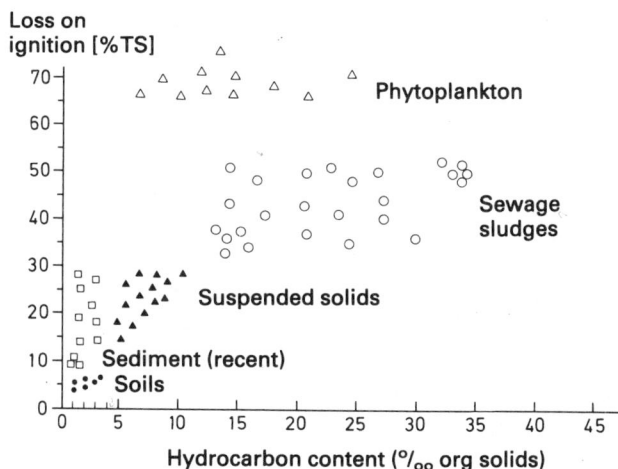

Fig. 8.13 — Hydrocarbon content of organic solids in aqueous media.

ideally starts from a chloroform extract. After transfer to CCl_4 one can obtain by IR spectrometry a value for the total extract, or in geogenic samples the 'bitumen' fraction. Following this the polar substances are removed according to the flow diagram contained in Fig. 8.14. Silica gel may be used in place of Al_2O_3 if so desired. The copper powder is chiefly for binding elemental sulphur, which can be present in fairly high concentration.

```
                        ┌──────────┐
                        │  Sample  │
                        └──────────┘
                             │
                     Extraction (CHCl₃)
                             │
                        ┌──────────┐
                        │ Extract  │
                        └──────────┘
                             │
                     Pretreatment (SiO₂)
                             │
                     Chromatography  (Column, Al₂O₃, Cu-powder,
                                        n-Hexane/Benzene 8:2)
                             │
                   ┌──────────────────┐
                   │ Extract, nonpolar │
                   └──────────────────┘
                             │
                     Chromatography  (Thin-layer Silica gel F-60
                                        n-Hexane/Benzene 8:2)
```

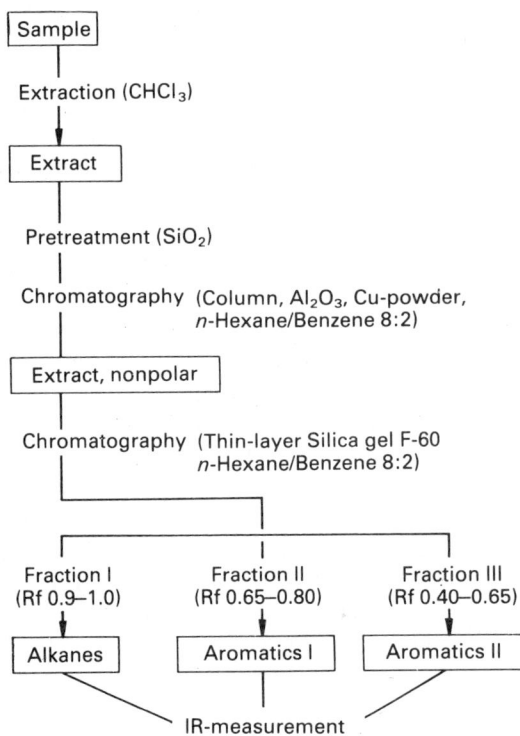

Fig. 8.14 — Preparation of a sample extract for hydrocarbon analysis.

In connection with the thin-layer chromatographic separation process, one isolates the group of alkanes plus, depending on the partition of the aromatics on the support, two groups of aromatics which differ in the numbers of aromatic nuclei. With the aid of IR spectra the composition of these groups can be quantitatively assessed. It is possible to distinguish in the IR spectra of Fig. 8.15, besides a high proportion of paraffin chains (2800–3000 cm^{-1}), the skeletal vibration of the aromatic nucleus at *ca.* 1600 cm^{-1} and the increasingly intense vibrations (from spectrum 2 to 3) between 700 and 900 cm^{-1} characteristic of polynuclear aromatics. If it is stated that in recently formed biogenic hydrocarbon mixtures, substances of the same chemical composition are absent, then of course the quantitative aspects must be duly considered.

One quite large problem consists in determining small amounts of mineral oil products in the presence of natural hydrocarbon compounds. So long as the latter predominate then the unsubstituted polycyclic aromatics are preponderant, as in numerous sludges, for example, from the West German ship canal. If mineral-oils preponderate, then the unsubstituted PAHs show a decrease in amount relative to the alkylated polycyclic derivatives, as can be seen from the gas chromatogram in Fig. 8.11 (lower).

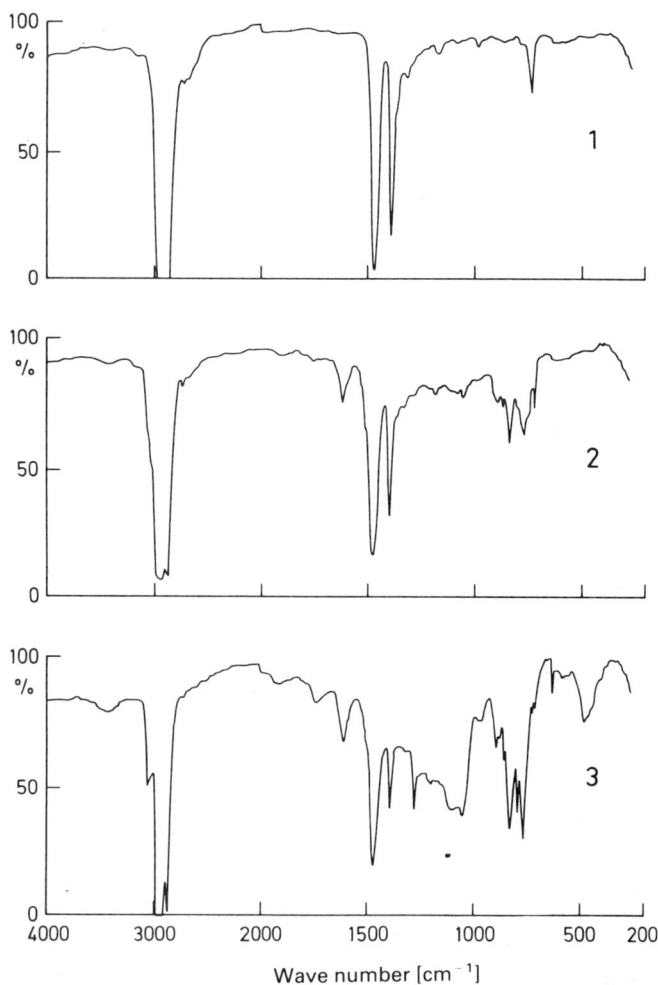

Fig. 8.15 — IR spectra of hydrocarbon fractions separated by thin-layer chromatography.
1 Alkanes; 2 Low-boiling aromatics; 3 Higher-boiling aromatics.

In concluding this section of many problems, a further example relating to the particle size question may be mentioned. According to Table 8.3, moisture contents and hydrocarbons concentrations in the dredge spoil samples are positively correlated. In addition the recorded heavy metal contents fit into the scheme. In place of the moisture content, loss on ignition and total extract are preferable as tracer parameters, as will be shown in the later chapters. The loss on ignition can serve for this purpose with more justification in the case of heavily polluted samples than the particle size distribution.

Table 8.3 — Organic matter and heavy metal contents in sediments from the Moselle, 1984

No.	Moisture content (%)	Sand fraction (%)	Hydro-carbons	PAHs	Pb (mg/kg)	Zn	Cr
1	30	94	640	9.7	169	642	21
2	33	95	890	9.2	180	606	22
3	26	97	810	11.6	173	633	22
4	66	14	4060	16.2	849	1787	171
5	63	17	3520	22.2	467	1215	128

LITERATURE

8.2

Hellmann, H. (1973). Mit welchen Analysenmethoden können Mineralölverschmutzungen dem Verursacher nachgewiesen werden?, Vom Wasser **41**, 45–46.

Scholl, F., Fuchs, H. (1968). Bestimmung von Mineralölspuren in Wasser, Bosch, Techn. Ber. **2**, 235–244.

Rübelt, C. (1968). Spurennachweis von Mineralölprodukten in Bodenproben mit der IR-Spektroskopie. Z. Anal. Chem. **221**, 299–304.

Kägler, S. H. (1969). Neue Mineralölanalyse, Dr. Alfred Hüthig Verlag, Heidelberg.

Hellmann, H. (1978). Einfache instrumentelle IR-Messung von Kohlenwasserstoffen in Fluß- und Abwässern, Vom Wasser, **50**, 231–246.

Hellmann, H. (1982). Fluorimetrie als Alternative zur IR-Spektroskopie bei der Kohlenwasserstoffbestimmung?, Vom Wasser **59**, 181–194.

Beurteilung und Behandlung von Mineralölschadensfällen im Hinblick auf den Grundwasserschutz, Teil 3; Analytik-Probenahme, Methodik, Interpretation, Hrsg. Umweltbundesamt, März 1979.

Anfärbereagenizien für Dünnschicht- und Papierchromatrographie, E. Merck, Darmstadt 1970.

8.3

Rübelt, C. (1970). Differenzierungen zwischen Mineralölprodukten und Kohlenwasserstoffen biogenen Ursprungs bei Mineralölschadensfällen. Fortschr. Wasser- und Ambwasseranalytik **231**, 65–78.

Hellmann, H. (1974). Unterscheidung von biogenen und mineralbürtigen Aromaten durch Fluoreszenzspektroskopie, Fresenius Z. Anal. Chem. **272**, 30–33.

Berthold, I. (1974). Analytik, Struktur und Vorkommen biogener Kohlenwasserstoffe. Teil II: Aromatische Kohlenwasserstofe. Erdöl und Kohle, Vorträge der 24. Haupttagung der DGMK, Ergänzungsband, 74–75, 940–951.

Hellmann, H. (1977). Analyse von Kohlenwasserstoffen im Rahmen des Gewässerschutzes, Vom Wasser **48**, 129–141.

Goebgen, H. G., Brockmann, J. (1977). Die Anwendung der Dünnschichtchromatographie in der Abwasseranalytik am Beispiel der Bestimmung schwerflüchtiger Mineralöl-Kohlenwasserstoffe, Vom Wasser **48**, 167–178.

Deutsche Gesellschaft für Mineralölwissenschaft und Kohlechemie e.V. DGMK-Projekt 150 (1977). Leitfaden zur Unterscheidung von biogenen und mineralölbürtigen Kohlenwasserstoffen, Hamburg.

Hellmann, H. (1978). Zur Unterscheidung von biogenen und mineralölbürtigen Kohlenwasserstoffen in Wasser- und Feststoffproben, Hydrochem. hydrogeol. Mit. **3**, 113–145.

8.4

Hellmann, H. (1973). Zum Abbau von Mineralölkohlenwasserstoffen im Rheinstrom, Tenside/Detergents **10**, 285–289.

8.5

Giger, W., Schaffner, C. (1975). Aliphatic, Olefinic and Aromatic Hydrocarbons in Recent Sediments of a Highly Eutropic Lake, Adv. Org. Geochem. 375–390.

Unger, U. (1971). Untersuchungen über die Verunreinigung des Bodensees durch Mineralöl, GWF **112**, 256–261.

Kölle, W., Stieglitz, L. (1974). Petrochemische und biogene Kohlenwasserstoffe in den Sedimenten des Bodensees, Vom Wasser **43**, 161–177.

Deutsche Gesellschaft für Mineralölwissenschaft und Kohlechemie e.V. DGMK-Bericht 294 (1982). Art und Herkunft der Kohlenwasserstoffe in Sedimenten des Bodensees, Hamburg.

Michael *et al.* (1975). Conference on Prevention and Control of Oil Pollution, zietiert nach Völtz, M., Ewers, H. (1983). Mineralölanalytische Aspekte zum Bodensee-Sediment, Erdöl, Kohle-Erdgas-Petrochemie **36**, 535.

8.6

Krejci-Graf, K. (1959) Diagnostik der Herkunft des Erdöls. I. Herkunft der Muttersubstanz des Erdöls, Umbildung der organischen Substanz, Erdöl und Kohl **12**, 706–712.

Hellmann, H. (1977). Zur Belastung von Sedimenten, Klärschlämmen, Luft- und Straßenstaub durch organische Stoffe, insbesondere Kohlenwasserstoffe und Phosphate, Vom Wasser **436**, 225–246.

Hellmann, H. (1975). Mineralölprodukte in den Westdeutschen Schiffahrtskanälen, Zeitschrift für Binnenschiffahrt und Wasserstraßen 2/1975, 48–54.

9

Polycyclic aromatic hydrocarbons

9.1 PRELIMINARY REMARKS

Polycyclic aromatic hydrocarbons (PAHs) are frequently treated, for both analytical and environmental questions, as a self-contained sub-group. On strictly theoretical grounds they belong to the aromatic hydrocarbons and hence they were previously considered to belong to the mineral-oil aromatics. Various aspects, however, indicate the need for special treatment of the PAH compounds. In the technical literature the term is understood to mean almost exclusively *unsubstituted* aromatics with three or more condensed nuclei. In contrast to this, however, the polynuclear aromatics present in mineral-oils consist chiefly of alkylated compounds, in the form of the more or less numerous alkyl isomers and as homologous series derived from the same polynuclear skeleton.

Surprisingly one encounters mostly the non-substituted compounds in natural waters and in sludges and sediments in which contamination with mineral-oils could possible have occurred. Hence the PAHs to be considered in this chapter are not necessarily to be equated directly with mineral-oil. They may indeed arise from mineral-oils, particularly as a result of incomplete combustion, just as from the combustion of coal and other organic matter. By reason of this mode of origin, the PAHs are occasionally termed 'pyrogenic aromatic hydrocarbons'.

Besides their formation as a product of pyrolysis, the question of a biogenic origin has been and still is the subject of discussion. Whether this involves chiefly synthesis by photo-assimilating organisms or bacteria, or a possible conversion of (unsaturated?) compounds in soils and sediments, still seems to be open to question.

Although PAHs are ubiquitously present in very small quantities and may be found in fresh plants as well as in recent sediments, on the other hand the high level of specific contamination in certain sludges, coke-oven effluents, runoff from heavily used roads, and atmospheric dust cannot be explained in terms of purely natural sources.

Besides their special chemical structure the carcinogenic properties attributed to certain polycyclic compounds provide sufficient warrant for treating them as a special class other than mineral-oil or hydrocarbon compounds.

9.2 GENERAL ANALYTICAL INFORMATION

According to the West German Drinking Water Ordinance dated 31 January 1975, the following six aromatic compounds must be determined: fluoranthene, 3,4-benzpyrene, 3,4-benzfluoranthene, 11,12-benzfluoranthene, 1,12-benzperylene, and indeno(1,2,3-c,d)pyrene. Experience has since shown that this selection by no means comprises all the PAHs in a given sample. However, numerous separate investigations confirm that, with certain exceptions, such a selection can be regarded as representative of the total family of PAHs. Of course it is also a matter of the particular problem in hand. Where a routine quality control of water samples is concerned (drinking water, river water or treated effluents), then such a *choice* limited to certain specific compounds can be justified. This approach was adopted by Kunte and Borneff (1976), among others, with the two-dimensional thin-layer chromatography procedure devised and perfected by Kunte in particular.

A more exhaustive recognition and identification of as many PAHs as possible is, however, the objective of mineral-oil analyses, such as those performed in the Federal Republic of Germany, mainly by Grimmer and coworkers. In this context they employ gas-chromatographic methods, by which it is possible to obtain a relatively complete picture of the make-up of the PAH fraction, in terms of individual compounds.

In recent times a further method has appeared in the shape of high-pressure liquid chromatography. By this means it is possible to combine a sharp separation, such as can be achieved in general using gas chromatography, with the very sensitive detection characteristics of fluorescence spectroscopy. Whereas, for example, the detection limit for individual compounds with gas chromatography using direct, non-split injection equipment lies around 0.1 μg, by means of fluorescence detection certain aromatic compounds may be detected in quantities of 0.001 μg or less. Hence for surface and drinking water samples where the concentrations of PAHs are invariably below 0.1 μg/l (and in individual samples are frequently below 0.01 μg/l), gas chromatography is less effective than the other two methods. If one cannot dispense with the wide-range performance capabilities of gas chromatography, then it will be necessary to start with between 50 and 100 litres of water. Working up such a large volume of water presents quite severe problems regarding the purity of the solvents and the analytical containers, and also creates further difficulties familiar to the experienced analyst in similar situations. Table 9.1 lists the individual PAHs detected with the aid of gas chromatography (packed column $KNO_3/NaNO_3$ eutectic) in a sediment extract, together with their relative proportions with respect to both the sum of the compounds listed and the total aromatic fraction, obtained from the total impulse number. The individual aromatic compounds in this list made up about 50% of the larger total, and of this about half again could be attributed to the six polycyclic compounds specified in the Drinking Water Ordinance. Errors of about $\pm 10\%$ of each measurement cannot be ruled out.

9.3 SPECIAL REMARKS ON ANALYSIS

The choice of the analytical method has necessarily to be related to the problem in hand. Where, for example, by means of a 'profile analysis' of the PAH fraction (i.e. the occurrence and relative proportions of the separate compounds) the origin of this

Table 9.1 — Proportion of individual aromatics in the total aromatic fraction of a sediment (packed column as in Fig. 9.7)

Compound	Fraction (%)	Fraction (%)[a]
Fluorene	2.7	1.6
Phenanthrene	12.1	7.0
Anthracene	5.7	3.3
Fluoranthene	9.3	5.4
Pyrene	6.4	3.7
Benz(1,2)anthracene	6.1	3.5
Chrysene/Triphenylene	8.3	4.8
Benzfluoroanthene	16.3	9.5
1,2-Benzpyrene	7.3	4.3
3,4-Benzpyrene	8.4	4.9
Perylene	2.1	1.2
Indenopyrene	9.4	5.4
Benzperylene	5.9	3.4
Total	100.0	58.1

[a]Relative to the total impulse count of the integrator.

class of compounds is to be further investigated (Müller, Grimmer and Böhnke, 1977), then gas chromatography, possibly in conjunction with mass spectrometry (GC/MS), is the only candidate.

One is thereby assuming that this 'profile' in the form of the spectral composition of the aromatic group of constituents, will not have undergone any appreciable change between the point of origin (e.g. coal combustion), including transport and retention in the atmosphere, and the final arrival in a natural water system. If one regards the polycyclic aromatics in the sediments of Lake Constance as pyrolysis products from the combustion of coal, then one should be alerted to this initially unproven assumption.

Quite a different state of affairs obtains in the exhaust gas or liquid effluent study at an industrial site, in which the concentration of a particular aromatic compound is subject to official limits, and the maintenance of the prescribed conditions is being monitored. Depending on the chemical structure of the material, high-pressure liquid chromatography with fluorescence detection can be most advantageous. In the analysis of drinking and process water supplies, the determination of the six aromatic compounds is performed under the provisions of the Drinking Water Ordinance (see Fig. 9.1).

According to DIN 38 409 Part 13 the use of two-dimensional chromatography with fluorescence detection is envisaged for this purpose. The aromatics are determined separately and their concentrations added together. The process is relatively tedious and hence high-pressure chromatography is frequently used

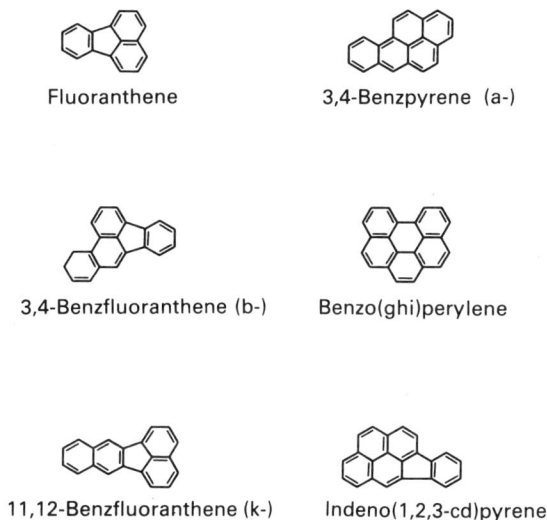

Fluoranthene   3,4-Benzpyrene (a-)

3,4-Benzfluoranthene (b-)   Benzo(ghi)perylene

11,12-Benzfluoranthene (k-)   Indeno(1,2,3-cd)pyrene

Fig. 9.1 — The six PAH compounds specified in the Drinking Water Ordinance.

instead. Besides this, semi-quantitative methods are also proposed; as long as the limiting value specified in the Ordinance (total of 6 PAHs = 0.25 $\mu g/l$ as carbon) is clearly not exceeded, a semi-quantitative test is regarded as sufficient.

It may also be allowable to mention a group method developed by the author, which—with the limitations basically attaching to all such group methods—permits quantitative predictions in the lower nanogram range, is simple to operate and gives a rapid answer. With the aid of this method it is also possible to detect the presence of mineral-oils or to confirm their absence.

First of all the position of the individual compounds on a thin layer composed of a mixture of aluminium oxide and acetylcellulose is shown in Fig. 9.2 following two-dimensional chromatography (Kunte). The six compounds specified in the Ordinance are shown hatched. In the intermediate stage, namely following uni-dimensional development of the extract with hexane/benzene (9:1) applied at a single point, one can discern under irradiation in the long-ultraviolet region three spots (Fig. 9.3, right) which correspond to compounds No. 2, Nos 7,8,9 and Nos 10,11 of Fig. 9.2.

If one chromatographs the same extract on silica gel F-60 instead, with the solvent hexane/benzene (8:2), then the spots 2 and 3 merge together. In the extracts obtained from water and more especially sediment samples, a third spot appears at lower R_f values (Fig. 9.3, left). These spots can, given suitable optical conditions, be quantitatively evaluated with the aid of a chromatogram spectrophotometer and integrated. If one selects a narrow-band excitation region at 365 nm and a broad-band emission with an edge filter K_f of 390 nm, then one obtains a chromatogram of the form shown in Fig. 9.4 (upper) for a sediment extract. The conditions for optimal fluorescence detection have been thoroughly substantiated (Hellmann 1979). As long as a calibration standard such as FER-A-POL is run in parallel, both fluoran-

1 Pyrene
2 Fluoranthene
3 Phenanthrene
4
5 Chrysene
6 Perylene
7 11,12-Benzfluoranthene
8 3,4 -Benzfluoranthene
9 3,4-Benzpyrene
10 1,12-Benzperylene
11 Indenopyrene

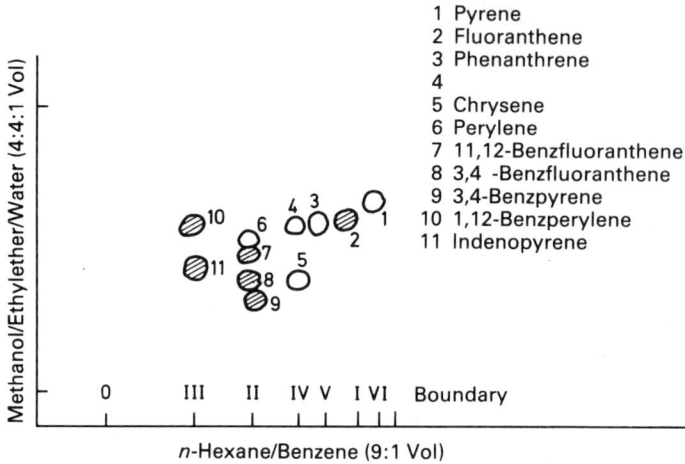

Fig. 9.2 — Positions of the PAHs after 2-dimensional chromatography on a thin-layer chromatographic plate of mixed Al_2O_3/acetyl cellulose.

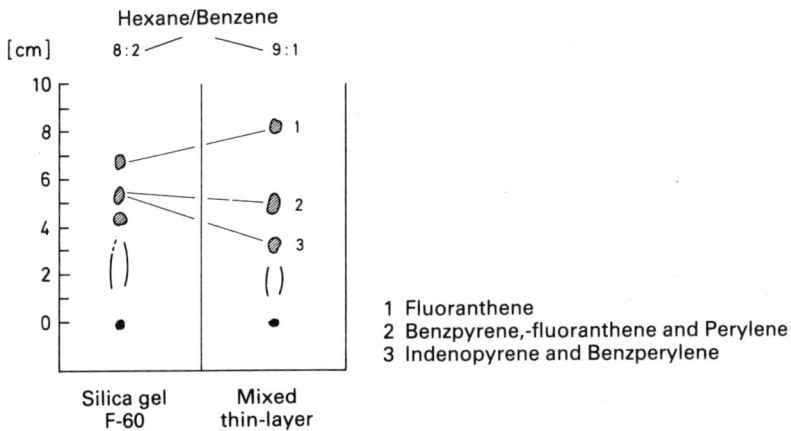

1 Fluoranthene
2 Benzpyrene,-fluoranthene and Perylene
3 Indenopyrene and Benzperylene

Fig. 9.3 — Comparison between resolution obtained on silica gel F-60 and the mixed thin layer adsorbent—1-dimensional development.

thene and the sum of the five remaining aromatics ($=\Sigma 5$) may be quantitatively determined.

With this mode of operation there appears on the chromatogram, in addition to the fluorescence peaks referred to, a field of greater or lesser extent which one terms the 'non-separable complex mixture' or diffuse background. In mineral-oil-free

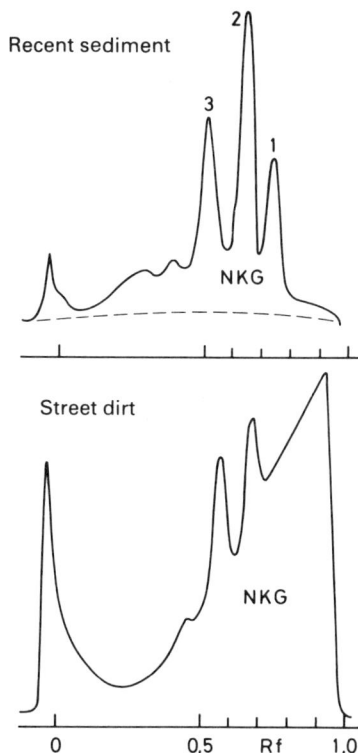

Fig. 9.4 — Fluorescence detection of extracts on silica gel F-60 using hexane/benzene (8:2), at 365/390 nm.

samples this background is quite small (Fig. 9.4, upper). The aromatic fraction characterised by fluorescence detection in Fig. 9.4 is also identical with the fraction analysed by gas chromatography according to Fig. 8.10 (lower). Conversely this background becomes dominant in the sample of street dirt heavily contaminated with mineral-oil (Fig. 9.4, lower), as it was in the equivalent gas chromatogram in Fig. 8.11.

Of course it is possible to operate with other optical settings, although no longer optimised with reference to the PAHs of the Drinking Water Ordinance. The combination 313/360 nm with narrow-band setting (Fig. 9.5) is well suited to the detection of typical mineral-oil aromatics, the alkylbenzenes, naphthalenes and phenanthrenes, etc. These can be determined with greater sensitivity in the short-wave than in the long-wave ultraviolet region. Even for mineral-oil-free sample extracts a chromatogram is obtained in the 313/360 nm wavelength mode, but its intensity is roughly half or less than half that of the analogous chromatogram with long-wave excitation. With appropriate experience it becomes possible to distinguish between samples obviously contaminated with mineral-oil and others which contain only very small amounts. Other examples will be found in Fig. 9.6.

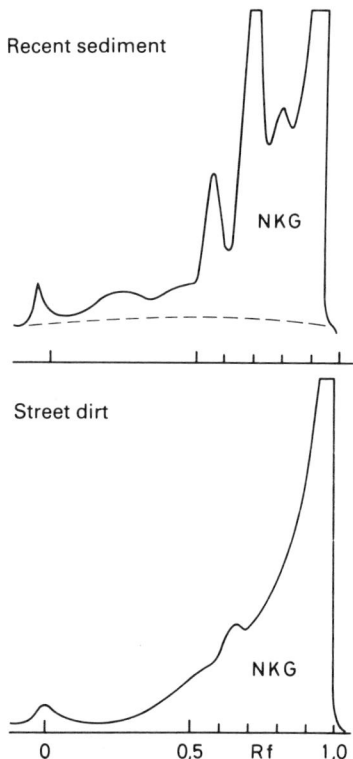

Fig. 9.5 — Fluorescence detection on silica-gel F-60 as for Fig. 9.4, but at 313/360 nm.

In the 'non-separable background' mentioned earlier are concealed—as in the case of alkanes—those biogenic materials the structures of which are not known or can be assigned only to certain bands of the chromatogram. Should additional mineral-oil aromatics be involved, an expert GC/MS analysis is likely to be the best solution.

Where larger quantities of extract are streaked on silica gel and developed with hexane/benzene (8:2), three fractions can be isolated with R_f values 0.65–0.80, 0.50–0.65 and 0.3–0.5, the gas chromatograms for which are illustrated in Fig. 9.7.

The first fraction contains as its principal components fluorene, phenanthrene/anthracene, pyrene and fluoranthene, of which only the last-named was covered by the group detection method in the 365/390 nm wavelength region. In the second fraction are found the basal constituents with five or six benzene nuclei, although traces of the four-member condensed ring compounds may be present. With regard to the third fraction, fluorescence on silica gel arises principally from the aromatics with seven or more benzene rings, while gas chromatography reveals a whole spectrum of lower to higher ring numbers.

From this example it can be seen that separation into zones in thin-layer plates is not always so clear-cut and that overlapping of particular compounds between

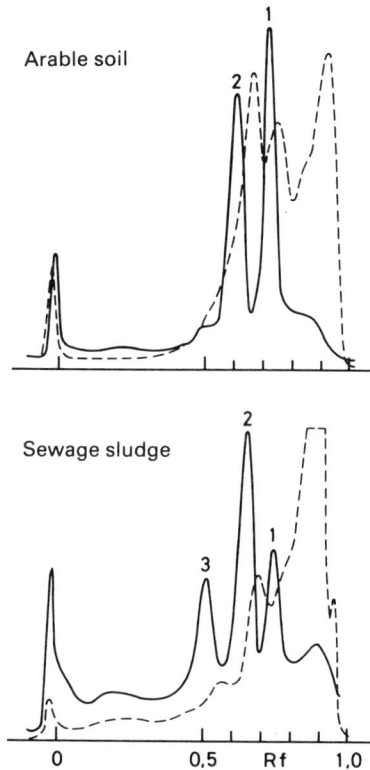

Fig. 9.6 — Fluorescence detection after development on silica gel F-60 as in Fig. 9.4. Wavelength combinations: 385/390 nm (full line) and 313/360 nm (dotted).

adjoining zones is not completely ruled out. For multiple mixtures or very heavy rates of application to the plate, this situation can be frequently observed and must not be forgotten, especially where quantitative determinations are concerned.

The PAHs in sediments and natural waters are apparently very persistent, yet during analysis considerable errors may be introduced on account of their volatility. The degree of volatilisation is relatively slight on the mixed-media thin layers referred to, but larger losses can occur in the case of silica gel. For this reason it has been suggested that operations should be conducted with paraffin-containing solvents (Hellmann 1979). One may usefully employ 2% paraffin in a mixture of *n*-hexane and benzene (9:1). By this means the adsorbent layer is to some degree impregnated and the fluorescence intensity increases by a factor of 10. The PAHs remain unchanged all day long, as long as one stores the thin layers in the dark. As a further bonus, the signal–noise ratio is improved (Fig. 9.8) and quantities of only 0.1 ng—with reference to the specific PAHs of the Drinking Water Ordinance—can be reproducibly and quantitatively determined. Where the individual compounds must still be determined separately, with the pressure aid of high-performance liquid chromatography (HPLC), the relevant portions of the adsorbent layer are lifted

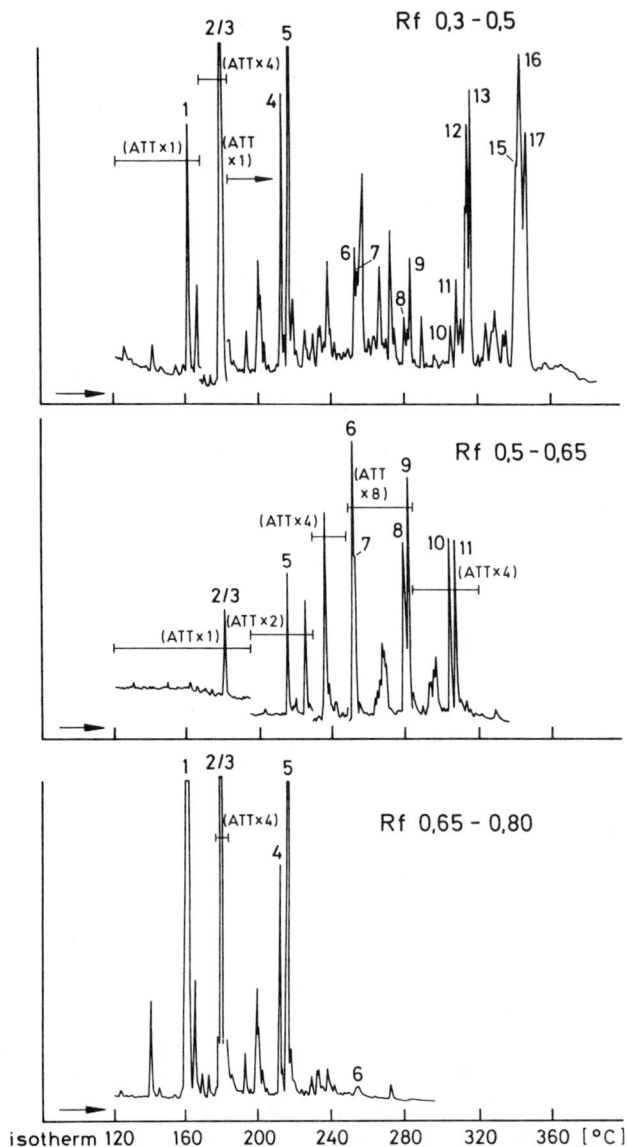

Fig. 9.7 — Gas chromatograph of aromatic fractions after separation on silica gel F-60 as in Fig. 9.4. Packed column K/NaNO$_3$ eutectic, 120–320 °C. *Key*: 1 Flurorene. 2/3 Phenanthrene/Anthracene. 4 Pyrene. 5 Fluoranthene. 6 Chrysene. 7 Triphenylene. 8 3,4-Benzpyrene. 9 3,4-Benzfluoranthrene (+ others). 10 Indenopyrene. 11 1,12-Benzperylene. 12,13 Dibenzofluor-anthrene (-anthracene). 14 —. 15 (?). 16 Coronene. 17 Dibenzo-pyrene.

away, powdered and eluted. The paraffin does not interfere with determination of the separate aromatics, but nevertheless ensures their persistence during the preparative treatment of the eluate.

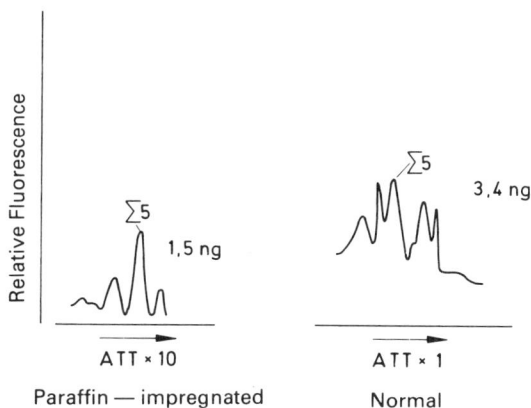

Fig. 9.8 — Comparison between fluorescence detection of PAHs on normal silica gel layers (right) and on silica gel after development with a paraffin-containing solvent (365/390 nm).

Comparative studies showed that the previously described group determination gave approximately the same numerical value of $\Sigma6$ as HPLC. Now and then the values obtained by the group method may be so high that the signal must be assumed to contain perylene and parts of the non-separable background among other things. Then the use of HPLC provides for somewhat better resolution of the background with advantage for the final result.

9.4   DISSOLVED AND UNDISSOLVED PAHS IN SURFACE WATERS

Although the limiting value for drinking water does not apply literally to surface waters, nevertheless it provides the surface water analyst with a point of reference for assessment of the results. In the case of the polycyclic aromatics, the Ordinance has also provided a guide as to the selection of those particular compounds for determination. Hence very often only the six compounds named in the Ordinance are determined. As can be seen from the relevant literature, the method most frequently employed is high-pressure liquid chromatography. As numerous results have been obtained certain regular features can be discerned, which can provide a guide for further interpretation and assessment.

First of all the composition of the dissolved PAHs at any point differs from that of the undissolved fraction. In the aqueous phase fluoranthene is by far the predominant constituent, accounting for 80–90% of the total, while in the suspended solids and sediments fluoranthene contributes only about 35%, or at most 50%, of the total. For the rest, the five other compounds are present in roughly equal amounts, although 3,4-benzfluoranthene is present in larger amounts than the 11,12-isomer.

Added to this in suspended solids-containing surface waters there is an unequal distribution of dissolved and undissolved PAHs, as shown by Fig. 9.9 with reference to a typical case of fluorescence detection on silica gel. The FER-A-POL reference material on the left of Fig. 9.9 contains 10 ng ($\approx$50%) of fluoranthene and

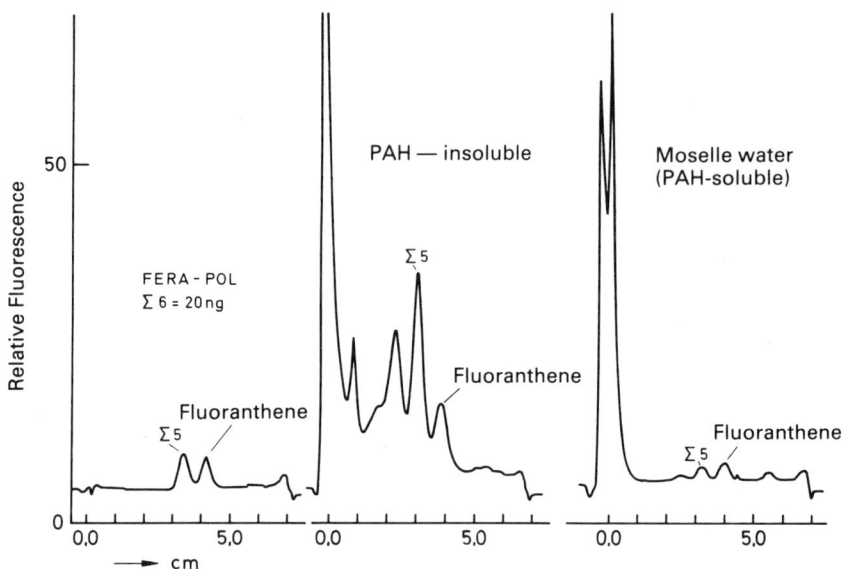

Fig. 9.9 — Determination of PAHs in Moselle water by means of fluorescence detection on silica gel F-60. Combination 365/390 nm.

2 ng ($\simeq$10%) of each of the other five polycyclic aromatics. In the centre the undissolved PAHs of a Moselle water sample are depicted as a chromatogram. The total quantity can be estimated at around 100 ng/l with fluoranthene representing about 30%. The concentration of dissolved PAHs (right) is obviously only about 10 ng/l, of which fluoranthene makes up more than 50%. For these measurements the full amplification of the spectrophotometer was far from completely utilised.

The quantitative determination of the undissolved fraction of the PAHs is to a certain extent bound up with the problem of the non-separable background component, as the Figure shows. Similar results are as a rule obtained for samples of water with suspended solids. The study of longitudinal river profiles reveals further differences in the occurrence of dissolved and undissolved PAHs. As a rule the concentration of dissolved aromatics changes very little (Fig. 9.10), the values lying well below the limiting value of the Drinking Water Ordinance. The downstream concentration change solely involves the undissolved PAH fraction. In order to assess the specific contamination of the suspended solids, the suspended solids content of the water must also be determined. Without these additional values it would be impossible to tell, for example, whether the concentration maximum in Fig. 9.10 for Speyer was the result of increased suspended solids *per se* or solids with a much higher level of contamination. If one also allows for the suspended solids concentration, an indication of the specific level of contamination of the suspended solids fraction is obtained (Fig. 9.11). For the samples obtained between 5 and 8 April 1982 it varied from 6 to 20 mg/kg, if one disregards the two peak values of 28 and 45 ppm. Both fluoranthene and the sum of the other five compounds ($\Sigma 5$) each

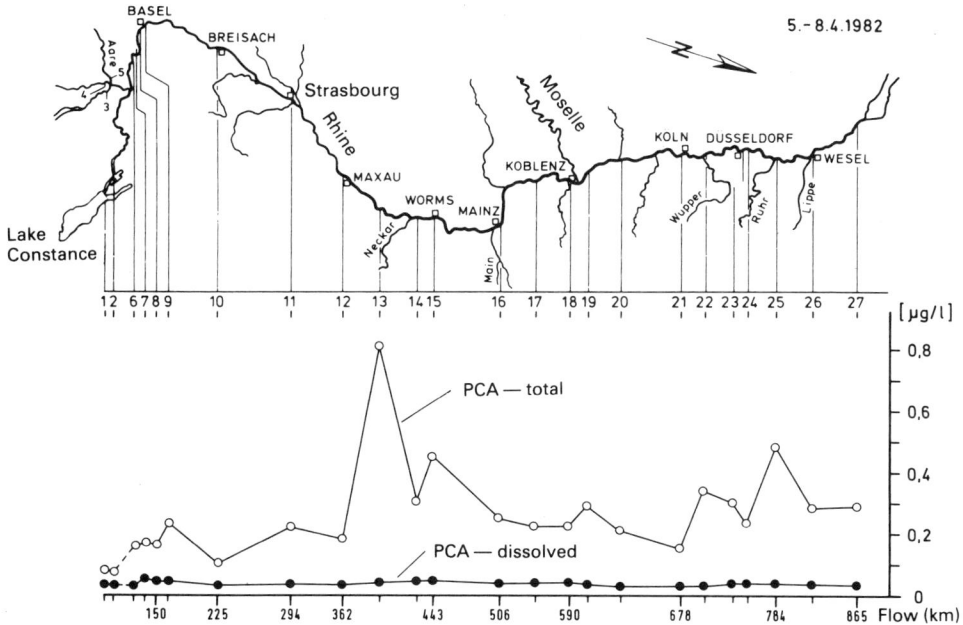

Fig. 9.10 — Concentration of dissolved and undissolved PAHs (as Σ6) along the length of the Rhine from Constance to Lobith. Sampling points as for Fig. 6.11.

constituted about 50% of the relevant amount. From the figure a further important point emerges. If one tolerates a certain range of scatter for the individual values, which is admissible on the grounds of unreliable measurements of suspended solids in the Upper Rhine, then the specific level of PAH contamination does not increase in the downstream direction, although the load of suspended solids of a roughly similar contamination level increases.

If one considers finally the ratio of fluoranthene to Σ5 PAH (Fig. 9.12), then no discernible changes in the direction of flow are apparent in respect of the *dissolved* PAHs. Fluoranthene invariably accounts for 80–90% of the total quantity. For the *undissolved* PAH fraction, however, the proportion of fluoranthene falls from around 70% in the Upper Rhine to around 50%. Since, as was shown, the concentration of fluoranthene in the water phase remains constant, this implies that discharges of substances lower in fluoranthene occur further downstream.

For the surface water analyst the following summary emerges: the concentration of dissolved PAHs specified in the DWO is ordinarily very low; it is below 0.05 μg/l and of this about 80–90% is composed of fluoranthene.

Of the suspended solids-associated undissolved PAH, 35–50% is composed of fluoranthene. The proportions of the remaining five polycyclic compounds in water and also in the suspended solids are relatively constant. As a result a lumped determination of Σ5 and also fluoranthene is normally adequate for purposes connected with hydrological investigations.

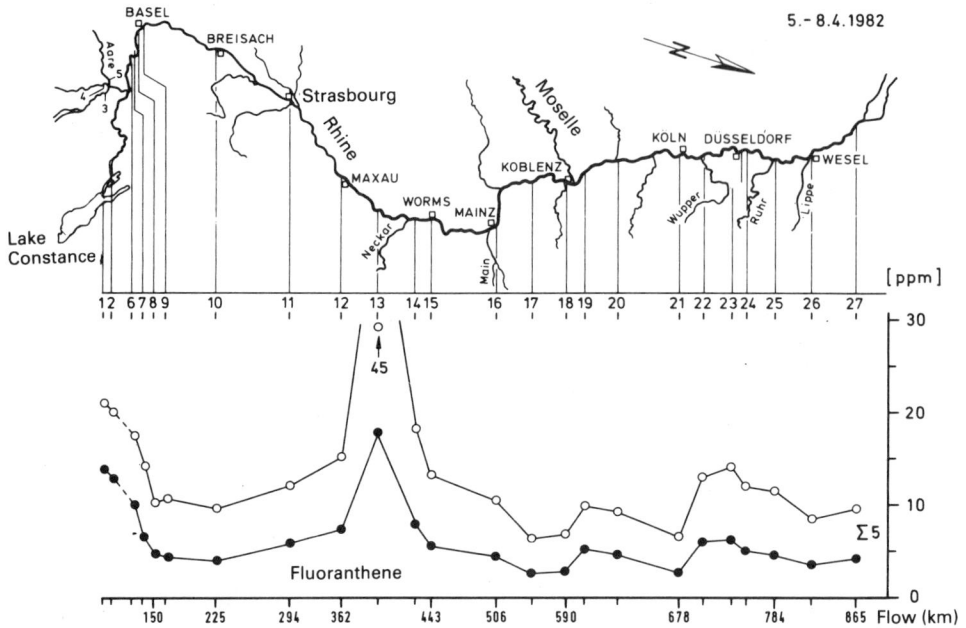

Fig. 9.11 — Specific PAH-contamination level of suspended solids in the Rhine—longitudinal profile.

Since the effects of discharge and variations in the discharge were dealt with in Chapter 7, in connection with heavy metals, together with the effect of the suspended solids transport on the results and their interpretation, these important factors will only be alluded to in passing here. Other aspects such as photochemical decomposition and, less important, biochemical degradation are still open to question.

9.5 PAHS IN SEDIMENTS AND SLUDGES

Suspended solids may be separated on a paper filter and dried in air in the dark, before the PAHs are extracted. The same can be done with sediment samples if they are adequately distributed in amounts of up to 0.5 g on a circular filter, For larger samples and those of higher moisture content, freeze-drying (Fig. 9.13) may be considered. This produces a porous-textured powder of good extractability. Where hot-air drying is employed the sample is frequently case-hardened or 'baked' so that the mass must first of all be mechanically comminuted before extraction—a not very desirable alternative. Relatively small starting weights of only a few grams may be readily reduced to a dry powder by triturating with anhydrous Na_2SO_4.

Just as individual PAHs in the pure form are susceptible to volatilisation and/or decomposition, so they appear equally resistant when enveloped in the solids matrix of natural waters or soils. All the same losses during either freeze-drying or hot-air drying cannot be ruled out *a priori*. Confirmation in this respect should be based on precautionary measures taken by the analyst. This applies equally to the investigation of hydrocarbons described in Chapter 8.

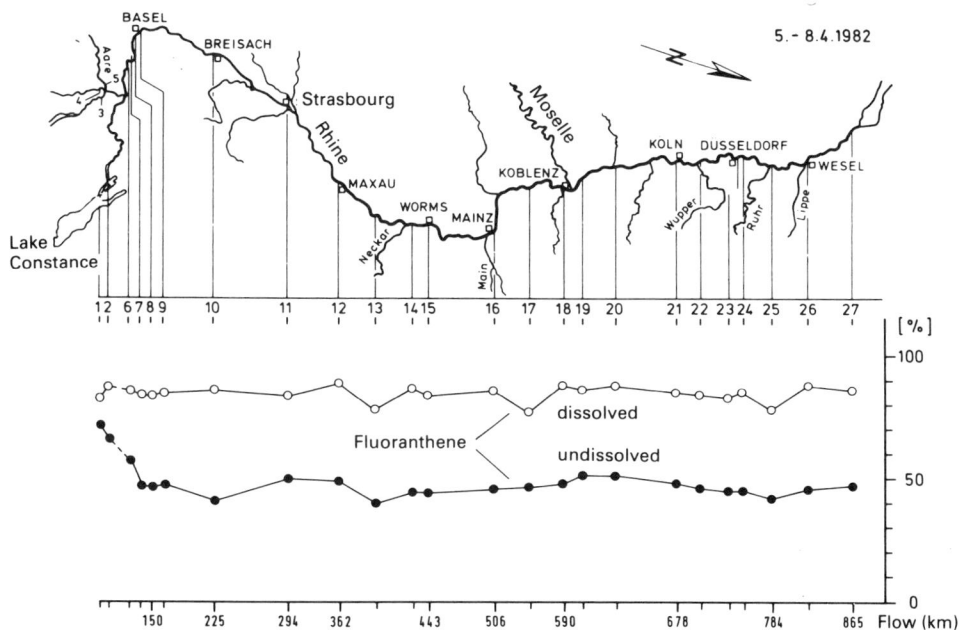

Fig. 9.12 — Fluoranthrene as a proportion of the DWO PAH compounds ($\Sigma5$) along the length of the Rhine.

The extraction of the dry sample with hexane/acetone, cyclohexane or chloroform, etc., may be performed in such a manner that the yield is as complete as possible, i.e. with the aid of the Soxhlet apparatus (Fig. 9.13). This is, however, seldom necessary. Almost always it is sufficient to take an aliquot of the extract, which is removed from the extraction vessel after settling of the solids. In this way the further step of centrifuging or filtration, together with the attendant sources of error, can be dispensed with.

Where fluorescence detection on thin-layer supports is the chosen method of detection, then microlitre volumes of concentrated extracts may be transferred directly onto the adsorbent plates. A further method may also be mentioned, according to which the sample is extracted directly, without pre-drying, with methanol or acetone in order to remove the aqueous components, and the PAH concentration determined graphically with the aid of solvent equilibrium diagrams (Hellmann 1984).

The flow chart presented in Fig. 9.14 is relevant to the determination not merely of PAHs but also of hydrocarbons, pesticides and PCBs from a *single* extract.

Concerning the problem of sampling of sediments and sludges, this has been dealt with in more detail in Chapters 7 and 8. One may expect, as for heavy metals and hydrocarbons, that the PAHs will be concentrated preferentially in the fine-particulate fraction, so that a correlation between the PAH content and the fine particulate fraction <0.02 mm in the samples from a single location may be observable. This fact is illustrated in Table 9.2, with the aid of a sediment sample

```
┌─────────────┐
│   Drying    │
└─────────────┘
    │
    ├──── Freeze-drying
    │
    ├──── Trituration with Na₂SO₄
    │
    ├──── Air-drying
    │
    └──── Thermal drying
```

```
┌─────────────┐
│  Extraction │
└─────────────┘
    │
    ├── Shaken with solvent (room temp.)
    │
    ├── Soxhlet (heat)
    │
    ├── Continuous extraction (column)
    │
    └── Stand in contact with solvent
```

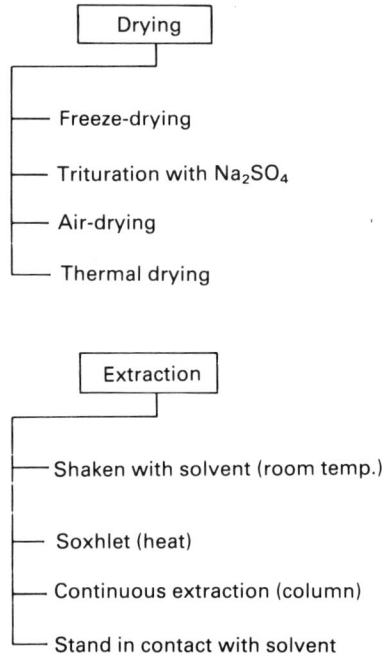

Fig. 9.13 — Analytical pretreatment alternatives for sludges and sediments.

sieved in an ultrasonic bath. Table 9.3 amplifies these results with reference to a non-fractionated sample of dredge spoil, for which, in addition to the loss on ignition, both the PO_4^{3-} content and the percentage of sand may be regarded as tracer parameters. As the loss on ignition will also include the coal dust fraction, this parameter is not very suited for correlation purposes, or use as a 'tracer'. This limitation is not applicable in Table 9.4, for which the samples were obtained from the regulated portion of the Neckar, at very similar sampling points in the impounded region, where they were deposited under nearly identical sedimentation conditions.

A wet-sieving step is not always necessary, as it can lead to errors as a result of discarding the particulate organic fraction. Thus one occasionally finds in recent sediments that for very low PAH contents ($\leqslant 1$ ppm) the measured PAH value is independent of the particle size of a dry-sieved sample, which is reflected in the PO_4^{3-} and loss on ignition (Table 9.5). In such cases the trace components are bound partly to the surface of the inorganic particles and partly to the particulate organic matter.

LITERATURE

9.1

Deutsche Gesellschaft für Mineralölwissenschaft und Kohlechemie e.V. (1974). Einfluß von Ottokraftstoffen auf die Emission von polynuklearen aromatischen Kohlenwasserstoffen in Automobilabgasen im Europa-Test, Forschungsbericht 4568, Hamburg.

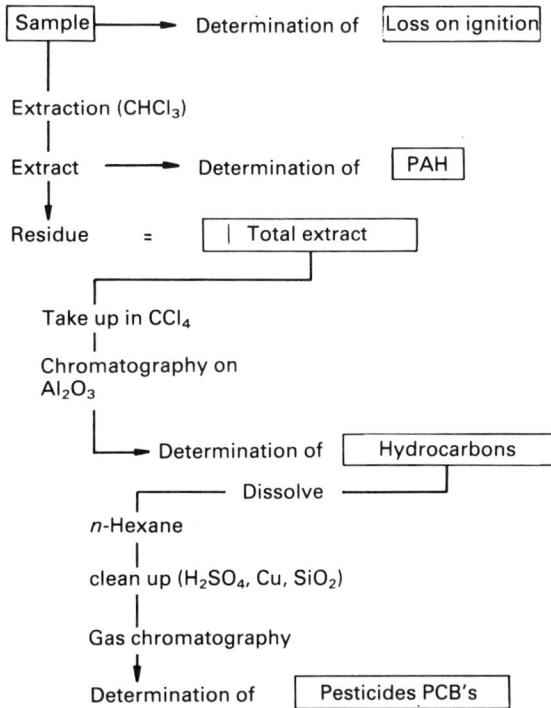

Fig. 9.14 — Flow-chart for determination of organics and trace substances in suspended solids, sludges and sediments.

Table 9.2 — Polycyclic Aromatic Hydrocarbons in a fractionated sediment sample (wet sieved in an ultrasonic bath)

Particle size distribution (mm)	Percentage fraction (%)	Loss on ignition (% DM)	PAH (mg/kg)
0.20–0.06	31	0.5	0.2
0.06–0.02	9	2.4	1.5
<0.02	60	19.3	14.1
	100	Av. 12[a]	8.7[a]

[a]Determined on the complete sample.

Table 9.3 — Polycyclic aromatics alongside tracer parameters in dredge spoil from the Moselle, 1982

Moisture content (%)	Percentage of sand (% DM)	Loss on ignition (% DM)	PO_4^{3-} (mg/kg)	PAH (mg/kg)
20	94	10[a]	780	9.7
26	97	7[a]	810	10.5
33	95	10[a]	960	10.8
63	17	15	1870	22.2
66	14	16	2390	24.5

[a]Representing coal dust.

Table 9.4 — Polycyclic aromatics alongside tracer parameters in sludges from the Neckar, May 1984

Sampling point	Loss on ignition (% DM)	$CHCl_3$ extract (mg/kg)	PAH (mg/kg)
Deizisau	20.2	4380	16
Obertürkheim	21.4	4270	9
Aldingen	18.5	4310	28
Poppenweiler	20.8	5040	10
Lauffen	16.5	4860	37
Gundelsheim	20.2	4500	25

Table 9.5 — Polycyclic aromatics in a recent sediment (after dry-sieving)

Particle size distribution (mm)	Loss on ignition (% DM)	PO_4^{3-} (mg/kg)	PAH (mg/kg)
1.0–0.5	16.6	3300	0.8
0.5–0.25	14.8	1260	0.8
0.25–0.125	8.5	1200	0.8
0.125–0.063	8.0	900	0.7
0.063–0.040	6.1	750	0.65
<0.040	6.0	840	0.75

Gräf, W., Diehl, H. (1966). Über den naturbedingten Normalpegel kanzerogener polycyclischer Aromate und seine Ursache, Arch. Hyg. **150**, 49–59.

Neff, J. M. (1979). Polycyclic aromatic hydrocarbons in the aquatic environment, Applied Science Publishers Ltd, London.

9.2

Kunte, H. (1967). Kanzerogene Substanzen in Wasser und Boden XVIII. Die Bestimmung von polycyclischen aromatischen Kohlenwasserstoffen mittels Misch-Dünnschichtchromatographie und Fluoreszenzmessung, Arch. Hyg. **151**, 193–201.

Kunte, H., Borneff, J. (1976). Nachweisverfahren für polycyclische aromatische Kohlenwasserstoffe im Wasser, Z. Wasser- und Abwasser-Forsch. **9**, 35–38.

Grimmer, G., Böhnke, H. (1976). Anreicherung und gaschromatographische Profil-Analyse der polycyclischen aromatischen Kohlenwasserstoffe in Schmieröl, Chromatographia **9**, 30–40.

Eisenbeiß, F., Hein, H., Jöst, R., Naundorf, G. (1977). Schnelle Trennung und Bestimmung von polycyclischen aromatischen Kohlenwasserstoffen in Wasser mit Hilfe der HPLC und integriertem Anreicherungsschritt, Chemie-Technik **6**, 227–231.

Althaus, W., Sörensen, O. (1971). Untersuchungen über den Gehalt der Vorfluter an krebserregenden Stoffen und die Beeinflussung dieses Gehaltes durch Abschwemmungen von bestimmten Straßenbelägen, Schriftenreihe der Deutschen Dokumentationszentrale Wasser e.V., H. 19, Düsseldorf.

Grimmer, G. (1972). Die quantitative Bestimmung von polycyclischen Aromaten mit der Kapillargaschromatographie, Erdöl Kohle **25**, 339–342.

Berthold, I. (1974). Einsatz der Hochdruckflüssigkeitschromatographie zur PNA-Bestimmung in Mineralölprodukten, Vorträge der 24. Haupttagung der DGMK, Compendium 74/75, 732–740.

Engelmann, H., Böhme, W. W. (1974). Bestimmung der polycyclischen aromatischen Kohlenwasserstoffe im Kondensat von Autoabgasen mit Hilfe der Flüssigkeitschromatographie, Vorträge der 24. Haupttagung der DGMK, Compendium 74/75, 741–749.

Hagemeier, H., Feierabend, R., Jäger, W. (1977). Bestimmung polycyclischer aromatischer Kohlenwasserstoffe in Wasser mittels Hochdruckflüssigkeitschromatographie, Z. Wasser- und Abwasser-Forsch. **10**, 99–104.

Berthold, I. (1974). Analytik, Struktur und Vorkommen biogener Kohlenwasserstoffe. Teil II. Aromatische Kohlenwasserstoffe, Vorträge der 24. Haupttagung der DGMK, Compendium 74/75, 940–951.

9.3

Müller, G., Grimmer, G., Böhnke, H. (1977). Sedimentary Record of Heavy Metals and Polycyclic Aromatic Hydrocarbons in Lake Constance, Naturwissenschaften **64**, 427–431.

Weil, L. (1980). Semiquantitativer Test zur Erfassung polycyclischer aromatischer Kohlenwasserstoffe im Trinkwasser, Z. Wasser Abwasser-Forsch. **13**, 108–111.

Stahl, E., Jork, H. (1968). Dünnschicht-Chromatographie; XIX Mitt.: Direktauswertung mit dem Chromatogramm-Spektralphotometer, Zeiss Inform. **16**, 52–59 (Nr. 68).

Hellmann, H. (1974). Unterscheidung von biogenen und mineralölbürtigen Aromaten durch Fluoreszenzspektroskopie, Z. Anal. Chem. **272**, 30–33.

Hellmann, H. (1975). Vereinfachte routinemäßige Bestimmung von polycyclischen Aromaten, Z. Anal. Chem. **275**, 109–113.

Hellmann, H. (1979). Optimierung eines fluorimetrischen Gruppennachweisverfahrens für polycyclische aromatische Kohlenwasserstoffe auf Kieselgel, Fresenius' Z. Anal. Chem. **295**, 388–392.

Hellmann, H. (1979). Zur Veränderung der Fluoreszenzintensität polycyclischer Aromaten auf Dünnschichtplatten, Fresenius' Z. Anal. Chem. **295**, 24–29.

9.4 and 9.5

Hagemaier, H., Kaut, H. (1981). Polycyclische aromatische Kohlenwasserstoffe in Sedimenten von Neckar, Neckarnebenflüssen, Rhein und Donau, Chem. Ztg. **105**, 181–186.

Grimmer, G., Hilge, G., Niemitz, W. (1980). Vergleich der polycyclischen aromatischen Kohlenwasserstoff-Profile von Klärschlammproben aus 25 Kläranlagen, Vom Wasser **54**, 255–272.

Fresenius, W., Fresenius, R. E., Schneider, W. (1977). Erste Ergebnisse der Trinkwasserüberwachung, Wasser, Luft Betr. **21**, 594–595.

Bundesanstalt für Gewässerkunde (1979). Organische Spurenverunreinigungen im Wass—Nachweis und Verhalten von polycyclischen aromatischen Kohlenwasserstoffen, Koblenz.

Bundesanstalt für Gewässerkunde (185). Organische Spurenstoffe in Sedimenten und Baggergut der Bundeswasserstraßen, BfG-0244, Koblenz.

Hellmann, H. (1983). Korngrößenverteilung und organische Spurenstoffe in Gewässersedimenten und Böden, Fresenius' Z. Anal. Chem. **316**, 286–289.

Hellmann, H. (1984). Remobilisierung und Bestimmung von k-Tensiden in Tonmineralien, Fresenius' Z. Anal. Chem. **319**, 270.

Jahresbericht ARW 1976.

Neff, J. M. (1979). Polycyclic aromatic hydrocarbons in the aquatic environment, Applied Science Publishers Ltd, London.

10

High-boiling organohalogen compounds

10.1 PRELIMINARY REMARKS

The chemically heterogenous group of pesticide compounds began to assume considerable importance in connection with surface water analysis during the 1960s, once again principally in the context of drinking water supply. With the development of analytical techniques and the consequent wide-ranging surveys of the concentrations of pesticides in natural waters and in the biosphere in general, the compounds which came to the fore were those which were evidently only poorly degradable (see also Thier and Frehse, 1986). Partly for this reason such compounds were able to accumulate in the solids present in natural waters and sediments, and thereafter in living organisms. They comprised chiefly the relatively non-polar organochlorine compounds, among which first of all the DDT group (o,p'-and p,p'-DDT and their metabolites DDE and DDD), followed by the isomers of hexachlorocyclohexane (α-, β-, γ-, δ-HCH) were analytically determined. To these were added the so-called 'drins' (aldrin, dieldrin and endrin) and also hexachlorobenzene, which until the middle of the 1970s was employed as a fungicide, but is now no longer used for this purpose.

While some contact between these materials and the water cycle, including effluent discharges and natural water systems, should have been seen to be inevitable, nevertheless this was not envisaged at the time, just as in the case of the polychlorinated biphenyls (PCBs) which came into prominence a little later. Nevertheless the PCBs along with the 'hard' organochlorine pesticides were shown to be present, with the aid of the continually improving analytical methodology, in natural waters and to be similarly concentrated in sludges and sediments. In very recent times the occurrence of polychlorinated triphenyls, dioxins and other chemical groups has also been reported.

In the meantime some countries, among them the Federal Republic of Germany, have implemented extensive usage restrictions, involving a ban on the production (DDT, PCB, 'drins') and marketing of such compounds. The global distribution of such compounds, as well as the quantities deposited in the sediments of natural waters, renders it advisable, however, to continue the determination of this group of halogenated organic compounds and to compile periodic accounts of their persistent residues.

Besides the nowadays recognised 'classical' halogenated organics, numerous other related compounds have latterly been shown to be present in surface waters, such as the chlorinated phenols, including pentachlorophenol (PCP), chlorinated benzenes, chloronitrobenzenes and other chlorinated benzene derivatives, and finally also some relatively volatile chlorinated solvents, from dichloromethane to perchloroethylene (Chapter 11). So far as one can tell at present, however, these groups of halogenated organics are not so important in aqueous media as the high-boiling compounds. Apart from their toxic properties and degradative behaviour, the accumulation potential of such low-boiling halogenated organics is also of a less serious nature.

10.2 BRIEF DETAILS OF ANALYSIS

The determination of high-boiling organohalogen compounds should be standardised. Even if in the foreseeable future a DIN document becomes available, however, their determination will remain one of the most demanding tasks of surface water analysis. Further to the already mentioned problem areas of sampling and — for solids — sample pretreatment, it is possible to distinguish three separate difficult and error-prone stages, viz. the (initially crude) separation of interfering and unnecessary materials, the so-called clean-up step; then the preliminary separation of the crude fraction into the various groups of relevant compounds; and lastly the final determination, and not infrequently the identification of the compounds by means of gas chromatography either alone or coupled with mass spectrometry (GC/MS).

As long as one is solely concerned with the identification of the individual organohalogens and if need be with the estimation of the relative amounts, e.g. of α- to γ-HCH, and sufficient sample is available, one can pursue the initial separation on columns or absorbent plates to the limit, as losses during any one operation will not affect the final result. The analysis becomes extremely demanding, however, where one is concerned with maximum permitted or limiting values in the sample, which may influence the fate or site of further retention of, e.g., awkward sludge deposits. To all appearances even the disposal of sludges from surface waters in future will be effectively controlled by analysis and by the limiting values obtaining at the time for specific chemicals.

The above problems are compounded, particularly for the quantitative determination of PCBs, by the necessary interpretation. As a rule only relatively few single components from the multitude of PCB isomers will be quantitatively determined. For this purpose one employs those isomers as tracers or indicator substances for which the chemical structure is accurately known, and which can be characterised by index numbers in the technical literature, such as those adopted by Ballschmiter *et al.* (1983) Then the question arises whether it is possible to scale-up to a total PCB figure, e.g. as Chlophene A-60. On purely pragmatic grounds this is desirable. Without going into the recognised ifs and buts, nevertheless the planned limiting values for sediments and suspended solids and also the financially important background may be commented on.

Apart from recommending the comprehensive specialist literature and the methods of analysis described in it, the presentation of a few chromatograms may serve to elucidate some of the topics referred to above.

In Fig. 10.1 it can be seen, for example, that the DDT group of compounds is not

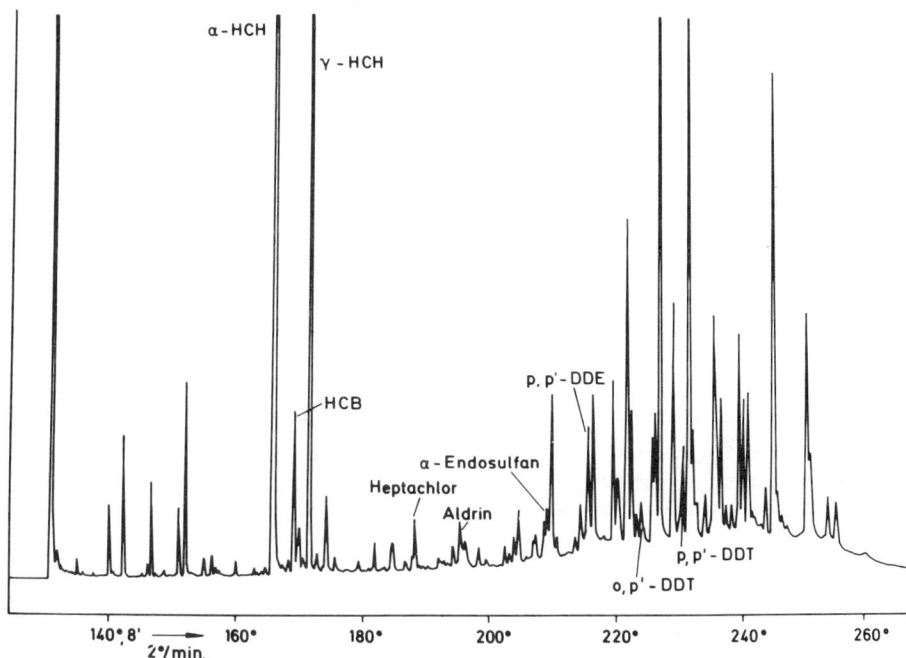

Fig. 10.1 — Organohalogen compounds in a sludge sample from the Neckar (1978). Glass capillary column OV101 (50m); 5' 140°C, 2°/min 140–240°C.

sufficiently distinguishable from the surroundings. The proposed clean-up method using Cu powder, concentrated sulphuric acid and column separation using silica gel F-60, in conjunction with the 50 m long glass capillary column, still proved inadequate. Instead of column chromatography, therefore, the more effective thin-layer chromatography technique may be employed. As can be seen from Fig. 10.2, the individual pesticides can be differentiated on the adsorbent plates to an astonishing degree. The question is, how far should the separation go in order to avoid interference during the succeeding gas-chromatography step? Using n-hexane, for example, the DDT group is adequately separated from the compounds in Chlophene A-60 (Fig. 10.3), the PCB fraction being split into three sub-groups. Over against the improved separation obtained by such a preliminary treatment, the user should also bear in mind the possibility of errors due to losses on the one hand and blank values from the adsorbent plate on the other. All the same these errors should be slight or at least predictable in the hands of a skilled and experienced analyst.

Certain results will be presented in the following sections, although the analysis is

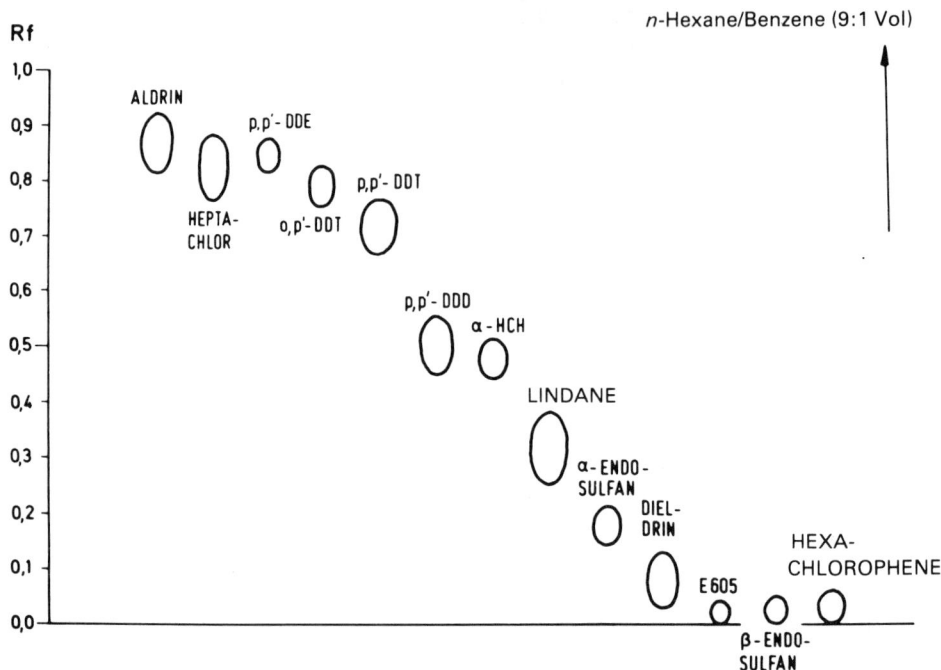

Fig. 10.2 — Separation of organochlorine compounds and other pesticides on silica gel
F-60 spray reagent Rhodamine B.

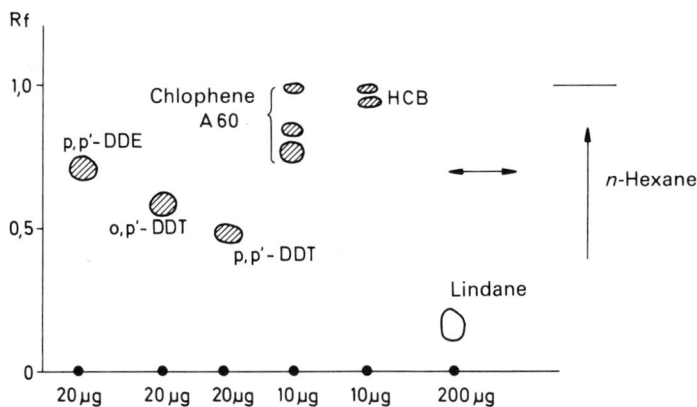

Fig. 10.3 — Separation of organochlorine pesticides and Chlophene A-60 on silica gel
F-60. Spray reagent Rhodamine B.

not by any means completely authenticated in every respect, for purely pedantic reasons. Thus certain publicised values are stated, but on the other hand the deductions which one may have to answer for are not ignored. Last, but not least, the hydraulic boundary conditions for plausibility checks should be included.

10.3 ORGANOHALOGEN COMPOUNDS IN RHINE SUSPENDED SOLIDS

For the compounds referred to in this chapter the following defines the distribution coefficient, for which values are of the order of 10^4–10^5:

$k = C_F/C_W$ (1/kg), where
C_F = concentration in suspended solids in mg/kg
C_W = concentration in water in mg/l

This means that the analyst has to manage with relatively small amounts of suspended solids. The following results relate to suspended solids obtained from 100 l of Rhine water by the settling process described in Section 5.2 (Table 10.1). The selection was made having regard to the water flowrate, which as can be seen ranged by almost an order of magnitude from 648 to 5650 m³/s while the associated suspended solids contents were in good accord with the hydraulic relationships outlined in Figs 2.1 and 2.2.

High discharge values accordingly implied elevated suspended solids contents and suspended solids fluxes which increased disproportionately with the discharge. The levels of α- and γ-HCH recorded in the suspended solids varied from 1 to 20 μg/kg (=ppb) on a dry wt basis, those for HCB from 30 to 200 ppb and for PCBs from 300 to 1100 ppb. If one considers first of all the PCBs then very high concentrations of around 1000 ppb and over are associated with relatively low suspended solids contents (Nos 7, 8, 10, 13 and 14) while relatively low PCB contents are associated with high suspended solids contents (Nos 11 and 16).

From this a relationship can be discerned between the suspended solids transport and their specific contamination level such as was indicated in Section 4.2 as well as in Figs 4.2 and 4.4 in general terms. For HCB the correlation appears less strict and for the two isomers of HCH it is not apparent from the values cited in the Table. Further deductions may be drawn if one transfers the data from tabulated to graphical form. The PCB mass flow then increases (Fig. 10.4) with the discharge, first from 600 to 2000 m³/s in a roughly linear fashion, and then from 2000 to 5650 m³/s disproportionately. Rapidly rising discharges generally create a scouring and eroding action, with disproportionate increases in suspended solids characterised by the 'pear-shaped' profile of the collective results shown in Fig. 2.2, in which the measured values frequently lie above or below the exact figures corresponding to an exponential function. This picture is reproduced by the HCB values (Fig. 10.5), with one outlier (?) in the shape of the sample for 2 December 1983, and possibly another in the shape of that for 20 January 1984.

The very similar pattern of occurrence of PCBs and HCBs in Rhine suspended solids at Koblenz (Fig. 10.6) is evidence on the one hand of their common origin and on the other of the 'internal consistency' of the analysis — which should not be confused with 'correctness'. Again in Fig. 10.6, the value for HCBs obtained on 2

Table 10.1 — Organohalogen compounds in suspended solids from the Rhine at Koblenz in 1983/84 (author's results)

Date	Flowrate (m³/s)	Suspended solids (mg/l)	Concentration in suspended solids (µg/kg DM)				Undissolved fraction in aqueous phase[b] (ng/l)	
			α-HCH	γ-HCH	HCB	PCB[a]	HCB	PCB
1　4 Nov.'83	789	34.5	6	12	13	520	0.4	18
2　11 Nov.'83	719	11	5	8	47	303	0.5	3
3　18 Nov.'83	648	10	21	10	63	301	0.6	3
4　2 Dec.'83	2120	56	13	5	507	369	16	21
5　9 Dec.'83	1110	13	17	6	126	575	1.6	7
6　27 Dec.'83	1910	35	5	2	135	670	4.7	23
7　6 Jan.'84	1460	15	7	1	160	1060	2	16
8　13 Jan.'84	1250	10	6	2	115	1150	1	12
9　17 Jan.'84	2930	126	5	2	76	555	10	70
10　31 Jan.'84	1840	21	3	6	92	950	2	20
11　10 Feb.'84	5650	150	11	3	37	290	6	44
12　17 Feb.'84	2140	83	5	4	110	470	9	39
13　28 Feb.'84	1360	12	24	13	94	1120	1	13
14　11 Mar.'84	1250	16	7	4	195	1350	3	22
15　6 Apr.'84	2310	29	9	4	59	465	2	13
16　15 Jun.'84	2020	80	14	6	35	280	3	22

[a]Calculated as Chlophene A-60.
[b]Contents of organohalogens in suspended solids converted to concentrations in the water phase.

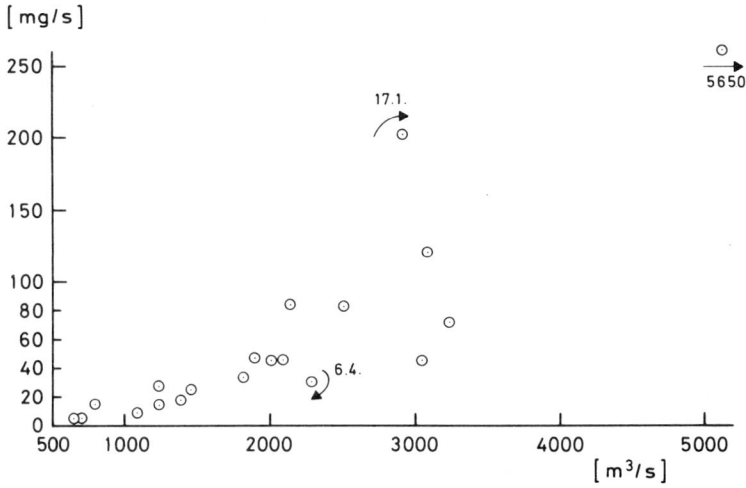

Fig. 10.4 — Relationship between PCB transport in suspended solids and discharge in the Rhine at Koblenz, 1983/84. Arrows show the direction of changes in the discharge.

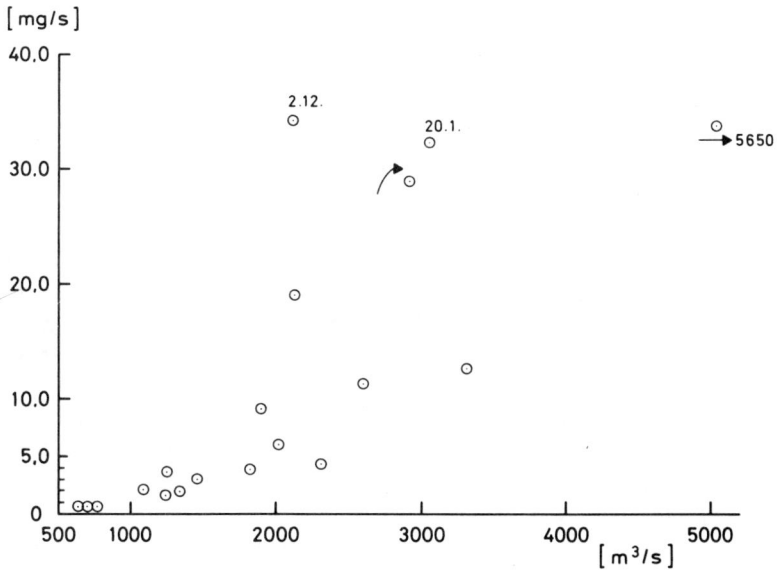

Fig. 10.5 — Relationship between HCB transport in suspended solids and water flowrate in the Rhine at Koblenz, 1983/84.

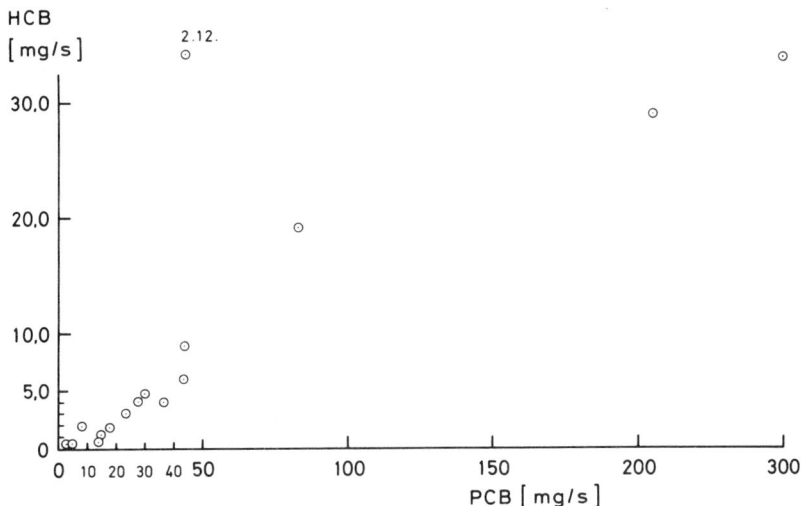

Fig. 10.6 — Relationship between HCB and PCB transport in suspended solids in the Rhine at Koblenz, 1983–84.

December 1983 is out of line. Since, however, very high values were also obtained in suspended solids on 29 November and 6 December 1983, the plausibility check does not provide any further information.

Apart from the values just referred to the concentration ratio of PCBs and HCB is approximately 5:1. If one assumes an average discharge value of 1560 m^3/s for Koblenz/Kaub as a basis for mass transport calculations, then on an annual basis the Rhine transports 950 kg PCBs and 180 kg HCB. The actual mass fluxes for these compounds are somewhat above these totals on account of the disproportionate increases during periods of flood flow.

The mass fluxes for α- and γ-HCH also increase with the discharge. The large scatter of the results for otherwise similar discharge conditions can probably be attributed to the less reliable analytical technique. The annual fluxes based on the MQ discharge are quite low, about 7 kg/annum for α-HCH and 4 kg/annum for γ-HCH. The ratio of 2:1 for these two isomers has also been reported elsewhere in the appropriate technical literature (data from the Elbe).

In conclusion the last two columns of Table 10.1 may be alluded to. There the values are converted to give the concentration of the respective substances as (insoluble) constituents of the water phase. The values for PCBs range from 0.4 to 16 ng/l and for PCBs from 3 to 44 ng/l.

Further information concerning the occurrence of halogenated organics can be derived from longitudinal profiles in the direction of flow. As an example values for the Rhine (Table 10.2) are cited. They belong, hydraulically speaking, to a moderate flood wave. The analyses were performed on 1-litre samples and the suspended solids were collected on a paper filter and not derived (as for Table 10.1) from settling of a larger volume of water. As expected from the rather adverse ratio of measured to background values, the errors are correspondingly large — Section 10.5. Under these conditions no undissolved organohalogens (PCB, HCB) could be detected in the Rhine at Constance with any degree of certainty, although some were probably present. At Öhningen (No. 2 in Table 10.2) the data must still be treated with

Table 10.2 — Organohalogen compounds in the Rhine–longitudinal profile measurements from 18 to 21 October 1982.[a] Undissolved constituents in the water phase

Point of sampling	HCB	PCB	HCB	PCB	Suspended solids	Distance
	(μg/kg)		(ng/l)		(mg/l)	(km)
1	40	1000	0.02	0.5	0.5(?)	0
2	50	125	0.02	0.5	4	
6	2	140	0.02	1.0	7	
7	3	125	0.02	1.0	10	
8	1	640	0.1	7.0	11	150
9	4	700	0.04	7.0	10	
10	8	670	0.04	5.5	5	225
11	130	500	0.2	7.0	14	294
12	130	428	0.2	6.0	14	362
13	110		0.2		16	
14	40	370	0.9	8.1	22	
15	35	370	0.8	8.6	23	443
16	300	390	7.0	8.6	22	506
17	110	330	3.3	9.6	29	
18	40	270	1.6	9.6	36	590
19	22	215	0.9	8.6	40	
20	18	270	0.6	9.1	34	
21	20	240	0.9	10.6	44	678
22	19	290	0.9	14.4	49	
23	19	300	0.9	14.6	48	
24	16	280	0.9	15.8	56	
25	30	440	2.0	29.3	66	784
26	50	585	2.8	32.2	55	
27	40	410	3.1	30.2	73	865

[a]Decreasing flood wave. For sampling points see Fig. 6.1

reservations. The specific contamination level in the Upper Rhine is accordingly very low. From Basle, however, the PCB contents of the suspended solids increased, even after conversion to the water phase. For HCB greater increases were observed at Kehl (No. 11) and Schierstein (No. 16). From Basle to Emmerich/Lobith the specific PCB contamination level of the suspended solids remained roughly constant, while for HCB localised peak values (effect of water flowrate?) and larger overall fluctuations are noticeable. Again after converting to the water phase, the PCB levels in the Rhine increased in the downstream direction from <0.5 to 30 ng/l, and those of HCB from <0.02 to 3.1 ng/l.

The difficulties in the evaluation of the gas chromatogram can be observed from Figs 10.7–10.9. The sample for Constance (Fig. 10.7) does not admit of any

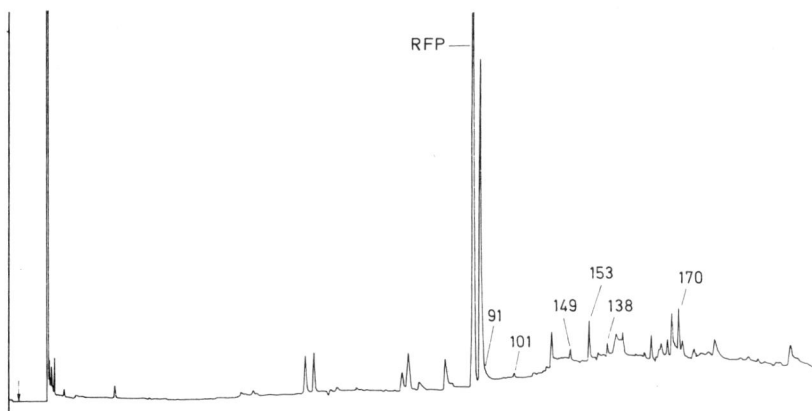

Fig. 10.7 — Gas chromatograph of suspended solids (1 litre) from the Rhine at Koblenz, 19 October 1982. Conditions as for Fig. 10.1.

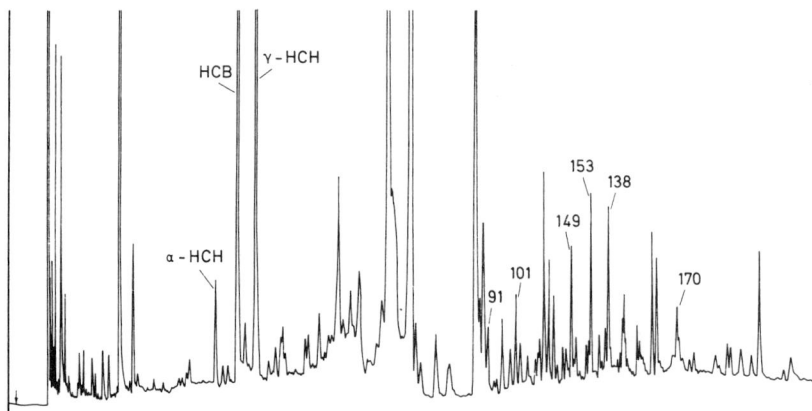

Fig. 10.8 — Gas chromatograph of suspended solids (1.0 l) in the Rhine at Emmerich/ Lobith, 21 October 1982.

deductions for either HCB or PCBs. At the German–Dutch frontier (Fig. 10.8), however, HCB is definitely measurable, while integration of the PCB peaks and in particular the ensuing scale-up to Chlophene A-60 presents problems.

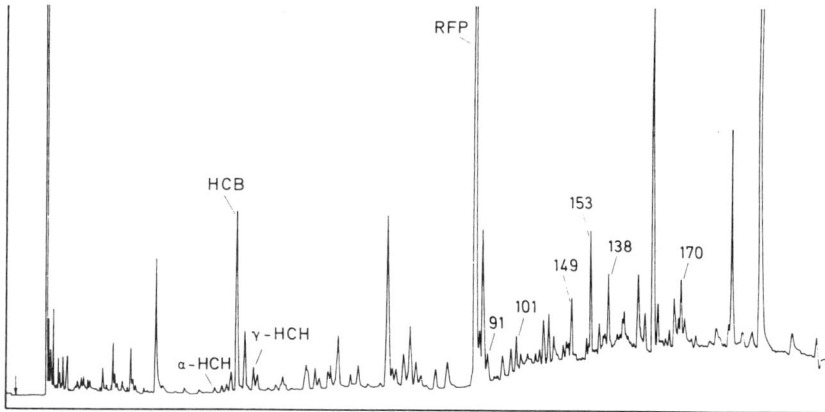

Fig. 10.9 — Gas chromatograph of suspended solids (*ca.* 10 l) in the Rhine at Emmerich/
Lobith, 8 November 1982.

The somewhat unsatisfactory limit of determination could be improved with the use of larger volumes of water. In the present case, the operation of the auto-sampler demands that the sample extract is diluted to 400 μl. Of the 400 μl only 1 μl is injected and this is then split. It is also worth mentioning that the peak measurements for the data in Table 10.2 were not performed using the on-line computer, which would have avoided the integration errors inherent in small signals.

10.4   ORGANOHALOGEN COMPOUNDS IN SLUDGES AND SEDIMENTS

Following the presentation in Table 10.1 and 10.2 of some contemporary data for organohalogen compounds in suspended solids from the Rhine, some analytical results for sludges will be considered. For a direct comparison between suspended solids and sludges the particle size effect and fractionation phenomena must necessarily be borne in mind. In the Oberwesel harbour region (Fig. 10.10), wide

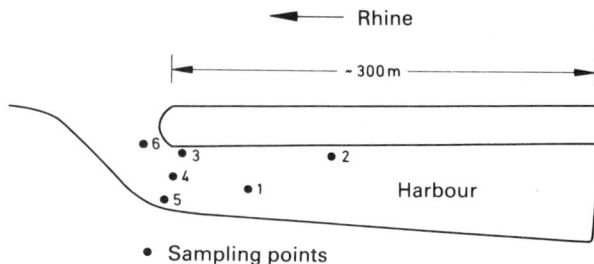

Fig. 10.10 — Location of sampling points in the harbour at Oberwesel.

variations were observed for different sampling positions, in line with expectations (Table 10.3). Particularly low HCB contents (0.5 ppb) were recorded at position 5, at

Table 10.3 — Organohalogen compounds in dredge spoil from the Oberwesel harbour (Middle Rhine 1981, corresponding to Fig. 10.10)

Sampling point	α-HCH	γ-HCH	HCB	o,p'-DDT	p,p'-DDT	PCB
			(μg/kg dry wt)			
1	50	8	50	1.5	4.7	600
2	47	8	67	1.2	3.2	680
3	59	9	139	0.9	3.3	510
4	70	10	96	1.0	3.3	730
5	3	1	0.5	0.3	1.1	270
6	41	7	33	0.6	2.6	300

which the hydraulic regime was unfavourable for sediment build-up. Very high HCB contents (139 ppb) occurred in the so-called current shadow (No. 3). The heavy metal contents at this point (not shown), and also the concentrations of phosphate and the values for loss on ignition which can be regarded as indicators, exhibited similar behaviour. Surprisingly high values — even compared to those for γ-HCH — were obtained for α-HCH in the sludge. This might call for a verification of the respective values by means of GC–MS analyses.

Frequently the analyst will not know the time at which the deposits were formed. To this extent any correlation with the polluting emisssion sequence is difficult to achieve. More helpful circumstances obtain in the case of inland lakes (Muller 1983) as well as possibly in certain tidal zones. Table 10.4 shows, for example, the results

Table 10.4 — Organohalogen compounds and associated parameters in a sediment core from the Ems, 1980

No.	Moisture content (% dry wt)	CHCl$_3$ extract (mg/kg)	Hydro-carbons (mg/kg)	PAH (mg/kg)	α-HCH	γ-HCH	HCB	PCB
							(μg/kg)	
1	66	2030	290	6.4	81	20	3.0	910
2	59	1970	250	5.2	72	15	19.0	720
3	65	1420	280	4.2	81	20	2.0	780
4	58	1425	240	4.8	70	16	2.0	400
5	66.5	3260	300	8.0	58	12	1.5	550
6	58	1190	240	4.3	72	12	1.5	630
7	61	1040	260	4.7	67	11	2.0	530
8	52	1035	210	3.3	54	9	1.5	500
9	60	1770	250	4.8	85	16	2.5	530
10	66	1450	290	6.1	100	18	2.0	525
11	60	740	250	4.9	61	16	2.5	910

for analysis of a sediment core from the Lower Ems. The individual segments can be dated with reference to their Cs-137 activity, a process analogous to that of the radiocarbon clock. From segment 4 downwards Cs-137 is no longer detectable, hence these layers are older than 20–25 years and originated prior to 1960–1965. The samples numbered 1–3 should therefore be assigned to the period 1965–1975. The most recent deposits are not represented in the Table.

If one considers first of all the PCB concentrations, these lie between 400 and 910 ppb. A time-related pattern is not detectable, and similarly for the other organohalogens and associated parameters. One may thus infer that over the decades material of roughly similar composition was deposited and the pollution situation remained roughly constant during this period. If one compares the individual values with those of Table 10.1 and 10.3 then the relatively high PCB content and very low HCB contamination of the segments of this depth profile are apparent. In contrast, the α-HCH and γ-HCH isomers are present in the sediment core in roughly equal amounts, as in the sludge dredged from the harbour at Oberwesel. At the same time the values are unmistakably several times greater than those of suspended solids from the Middle Rhine in 1983/84 (Table 10.1). Consequently it appears that the level of pollution of solids in the Rhine, in terms of these compounds, has markedly decreased.

As a tentative conclusion one may deduce that along with progressive PCB contamination of certain rivers and river reaches, HCB occurs preferentially, and that the level of HCH-isomers in general diminishes sharply.

10.5 ORGANOHALOGEN COMPOUNDS IN WATER SAMPLES FROM THE RHINE

As already mentioned, the exact determination of organohalogens assumes a certain definite signal/noise or signal/interfering peak ratio. If very small sample quantities are employed in conjunction with an ECD detection system, the values given by the computer on the print-out must be treated with a measure of scepticism (cf Section 10.4). Just such a situation is illustrated by the results of Table 10.5. They were obtained by working up samples of 1.0 l of Rhine water in each case, which was filtered through a paper filter to remove suspended solids. The water and the filter were then analysed separately.

If one compares the values with respect to organohalogens with those of Table 10.1 then the two sets of results are seen to be at variance although of the same general order of magnitude. Thus the concentration of α-HCH in Table 10.5 is clearly lower than the results shown in Table 10.1, the values for γ-HCH are high, and by comparison with α-HCH definitely too high. The HCB values are mostly too low and the PCB values rather too high. Completely unreal values are also apparent for the PCB-HCB ratio, ranging from 0.2 to 580:1 (right-hand column in Table 10.5) compared with a ratio of 5:1 obtained in Table 10.1.

Understandably such comparisons are accompanied by certain reservations as the samples were taken 2–3 years apart. However, a plausibility check with reference to the hydraulic situation also proves unsatisfactory. Thus for high discharge values and high suspended solids contents, high organohalogen contents (ng/l) should be obtained for the undissolved fraction with diminished organohalogen contents for

Table 10.5 — Organohalogen compounds in the Rhine at Koblenz 1981 (author's results)

Sample date	Discharge (m³/s)	Suspended solids (mg/l)	HCH sol.	HCH ins.	HCH sol.	HCH ins.	HCB sol. (ng/l)	HCB ins.	PCB sol.	PCB ins.	PCB/HCB insol.
5 Jan.	2870	138	1	0.5	14	4	2	3	14	41	14
2 Feb.	1250	90	2	0.1	9	3	5	0.1	11	58	580
3 Mar.	1260	20	2	0.2	12	6	8	0.4	64	56	140
30 Mar.	2980	37	2	0.2	9	4	4	17	51	39	2
27 Apr.	1600	23	4	0.3	20	10	39	0.3	59	69	230
25 May.	1440	18	7	0.2	21	9	5	0.3	101	39	130
22 Jun.	1780	10	7	0.1	8	1	6	0.1	40	26	260
20 Jul.	1690	23	7	0.2	16	3	8	0.6	30	30	50
17 Aug.	1750	25	3	0.1	6	1	6	0.1	32	24	240
14 Sep.	1220	14	7	0.1	14	8	11	0.6	41	30	33
12 Oct.	2130	21	3	11(?)	10	6	8	37	770	6	0.2
9 Nov.	2210	29	9	7	25	8	19	6	15	6	1
7 Dec.	2890	63	13	10	35	7	11	5	22	7	1

the water phase. No evidence of this can be seen in Table 10.5. In addition the specific contamination level for the suspended solids shows fluctuations outside the normal range. In short the data given in Table 10.5 provide a forceful indication of the need to check the analytical technique, particularly the measurement instrumentation.

LITERATURE

10.1

Thier, H.-P., Frehse, H. (1986). Rückstandsanalytik von Pflanzenschutzmitteln. Analytische Chemie für die Praxis, Georg Thieme Verlag, Stuttgart.

Quentin, K.-E., Huschenbeth, E. (1968). Das Pestizidproblem als aktuelles Problem der Wasseranalytik und Wasseraufbereitung, Vom Wasser **35**, 76–94.

Gewässer und Pestizide, Schriftenreihe des Vereins für Wasser-, Boden- und Lufthygiene Nr. 34(1971) Gustav Fischer Verlag, Stuttgart.

10.2

Laubereau, P. G. (1984). Zum Stand der Normung eines Verfahrens für die Bestimmung von schwerflüchtigen Halogenkohlenwasserstoffen und Organochlorpestiziden in Wasser, Korrespondenz Abwasser **31**, 499–508.

Schulte, E., Acker, L. (1974). Gas-Chromatographie mit Glaskapillarsäulen bei Temperaturen bis zu 320°C und ihre Anwendung zur Trennung von Polychlorbiphenylen, Z. Anal. Chem. **268**, 260–267.

Buchert, H., Bihler, S., Ballschmiter, K. (1982). Hoch auflösende Gaschromatographie persistenter Chlorkohlenwasserstoffe (CKW) und Polyaromaten (AKW) in limnischen Sedimenten unterschiedlicher Belastung, Fresenius' Z. Anal. Chem. **313**, 1–20.

Ballschmiter, K., Buchert, H., Scholz, C., Zell, M. (1983). Baseline Studies of the Global Pollution. VI. The Pattern of Pollution by Chlorinated Hydrocarbons in the Caspian Sea, Fresenius' Z. Anal. Chem. **316**, 242–246 (und weitere Arbeiten der Verf.).

Malisch, R. (1971) Sedimente als Modelle für die Beurteilung der Umweltkontamination durch chlororganische Pestizide, polychlorierte Biphenyle und Phthalate unter besonderer Berücksichtigung des zeitlichen Verlaufs, Dissertation Münster.

Schulte, E., Malisch, R. (1983). Berechnung der wahren PCB-Gehalte in Umweltproben, I. Ermittlung der Zusammensetzung zweier technischer PCB-Gemische, Fresenius' Z. Anal. Chem. **314**, 545–551.

Hellmann, H. (1981). Persistente organische Schadstoffe in Gewässern und die Verfeinerung ihres Nachweises im letzten Jahrzehnt, Dtsch. Gewässerkd. Mitt. **25**, 114–119.

10.3–10.5

Arbeitsgemeinschaft für die Reinhaltung der Elbe (1983). Chlorierte Kohlenwasserstoffe — Daten der Elbe — 1980–1982, Hamburg.

Malisch, R., Schulte, E., Acker, L. (1981). Chlororganische Pestizide, polychlorierte Biphenyle und Phthalate in Sedimenten aus Rhein und Ruhr, Chem. Ztg. **105**, 187–194.

Müller, G. (1983). Zur Chronologie des Schadstoffeintrags in Gewässer, Geowissenschaften in unserer Zeit, **1**, 2–11.

Bundesanstalt für Gewässerkunde (1982). Entwicklung der Analytik zur Bestimmung von Pestiziden und polychlorierten Biphenylen, BfG-0015.

Bundesanstalt für Gewässerkunde (1985). Organische Spurenstoffe in Sedimenten und Baggergut der Bundeswasserstraßen, BfG-0244.

Borneff, J., Harmetz, G., Fischer, A. (1978). Die Belastung von Oberflächenwasser mit Chlorbenzolen, Forschungsbericht 14–74 des Hygiene-Instituts der Universität Mainz.

Bayerische Landesanstalt für Wasserforschung (1985). Organische Schadstoffe in bayerischen Vorflutern, München.

Landesanstalt für Wasser und Abfall Nordrhein-Westfalen: Verschiedene Berichte und Jahresberichte, Düsseldorf.

Bericht der IAWR 1979 und Jahresbericht der ARW 1976.

Fischer, A., Slemrova, J. (1978). Die Belastung des Rheins mit Chlorbenzolen, Vom Wasser **51**, 33–46.

11

Low-boiling organohalogen compounds

11.1 PRELIMINARY REMARKS

On 22 June 1982, 129 individual substances were selected by the Commission of the EEC as meriting further investigation. This proposal was based on the EEC Directive 76/464 for the 'Protection of Surface Waters from Contamination by the Discharge of Injurious Materials'. Besides the toxic metals mercury and cadmium, the high-boiling organohalogen compounds DDT, HCH, HCB, and PCBs and PAHs which were discussed in the preceding chapter, and some low-boiling organochlorine compounds, were also included in the EEC list. These comprised, among others, dichloromethane (No. 62), 1,1,1-trichloroethane (119), chloroform (23), carbon tetrachloride (13), trichloroethylene (121) and tetrachloroethylene (111). Further compounds of similar composition, such as the isomers dichloro-ethane and dichloroethylene, dibromomethane and tetrachloroethane which were also included in the list, will not be considered in the following paragraphs.

Chloro- and bromo-derivatives of methane (haloforms) were and are chiefly detected in certain drinking waters, and less so in sewage and effluents. In surface waters no bromine derivatives analogous to the low-boiling chlorine derivatives are observed. Despite the widespread distribution of chloroform, tri- and perchloro-ethylene, they do not present any immediate problem for surface water systems in the German Federal Republic since they are present in only very low concentrations in the lower microgram/litre range and frequently in the nanogram/litre region. Their very slight tendency to accumulate in the solids in natural waters must also be considered beneficial; moreover, this particular group is by no means comparable in persistence with the high-boiling group such as DDT and PCBs. If therefore the potential risk applies more to groundwaters than to surface waters (Leitfaden 1983), these should be the focus of analytical investigations. Besides the particular objec-tives of the EEC Directive 76/464, it nevertheless does not seem superfluous to devote some effort to the determination of low-boiling chlorinated hydrocarbons, at least occasionally in surface waters from certain districts. For on the one hand the major proportion of the discharge at certain seasons is derived from the groundwater flow, while on the other hand considerable concentrations of specific substances may arise in smaller streams as a consequence of point-source discharges.

11.2 ANALYSIS OF LOW-BOILING ORGANOCHLORINE COMPOUNDS

Prior to their ultimate determination, an enrichment process for chlorinated organic hydrocarbons from the water phase may be performed by three different methods, either liquid–liquid extraction, or the so-called headspace process, or stripping. Since the method of liquid–liquid extraction has been standardised the other two methods are of lesser importance in West Germany, although in the author's opinion the method of headspace analysis has proved to be very suited to the particular task of determining chlorinated hydrocarbons in surface waters (Hellmann 1985a, see Table 11. 1). The time and effort required is minimal and the results

Table 11.1 — Allocation of retention time (min) and particular compounds

Retention time, packed column (min)	Peak designation— specific compounds (after Hellmann 1985)	Retention time, capillary column (min)
0.79	Airpeak (pressure pulse?)	5.21
1.26	1,1-dichloroethane	5.38
	Freons?	6.09
1.44	Dichloromethane	6.41
	1,1,1- and 1,1,2-tri- chlorotrifluoroethane	6.54
1.96	Trichloromethane	7.31
2.31	1,1,1-trichloroethane	7.97
2.55	Tetrachloromethane	8.42
2.99	Trichloroethylene	9.13
?	1,1,2-trichloroethane	9.41
5.65	Perchloroethylene	13.90

adequately reproducible. Depending on the separating column, different individual compounds may be identified. With packed columns (e.g. 0V 101) the determination of tri- and tetra-chloromethane and also tri- and perchloroethylene presents no difficulties (Fig. 11.1). With high-resolution capillary columns (OV 1) some additional compounds may also be dealt with (Fig. 11.2). Whether this or that compound can still be detected and quantitatively estimated depends, among other things, on their very specific detection sensitivities—at equal concentration levels. Normally, in surface waters chlorinated organics with at least two chlorine atoms in the molecule, as in dichloromethane, are first detected. Using the headspace process in conjunction with ECD, the minimum detectable concentration for dichloromethane is in our experience around 0.1 mg/l, which is a concentration that may from time to time be encountered in the vicinity of sewage or effluent outfalls.

In studies using packed columns (OV 101) dichloromethane and the two trichlorotrifluoroethanes appear at the same point with identical retention times;

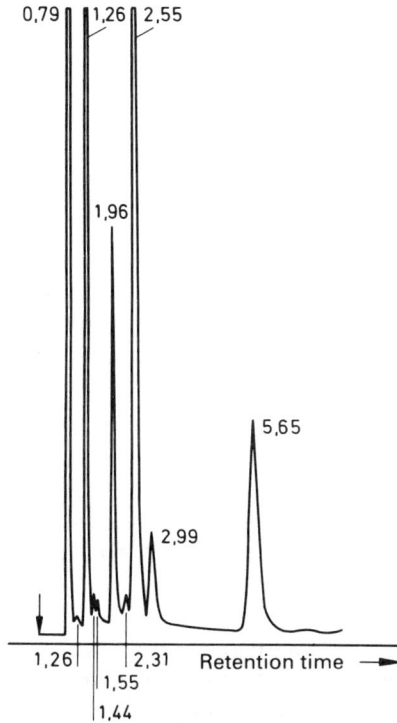

Fig. 11.1 — Chlorinated hydrocarbons in the Lower Main (27 March 1984); packed column 2 m (OV101). Peak identification as in Table 11.1.

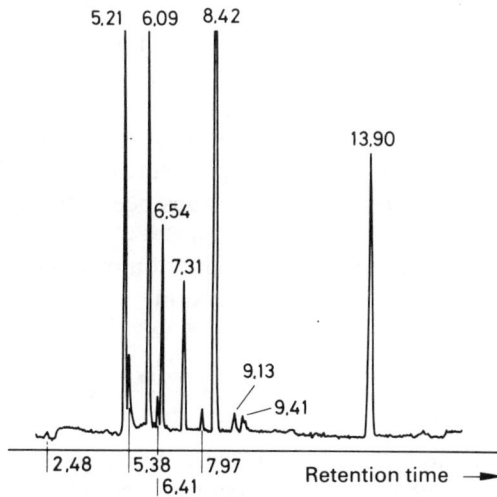

Fig. 11.2 — Chlorinated hydrocarbons in the Lower Main (28 January 1985); 30 m capillary column (OV1). Peak identification as in Table 11.1.

they are thus, for identical column coatings, only distinguishable using capillary columns. Trichlorotrifluoroethane is, however, scarcely found in surface waters. Prior to dichloromethane there may appear peaks due to freons (e.g. R 11 and R 13) in samples from certain rivers, which can only be satisfactorily determined using suitably coated capillary columns.

Besides the three enrichment processes referred to, a further method is also envisaged, in which the water samples can be directly injected onto the column (Schulz 1984).

A particular problem is presented by the analysis of river sludges, where one is unable to dispense with an analysis, for the reasons indicated below.

According to experience to date, the specific pollution due to the four chlorinated organics from trichloromethane (chloroform) to perchloroethylene is in the region of ≤ 0.1 mg/kg. The generally highly practical headspace technique is then only applicable with reservations. As long as original sludge samples are used in the sample containers, interruptions are likely to occur in the injection system. The obvious alternative of dispersing the sludge in organochlorine-free water and after equilibration, to analyse the sample, has met with some success (Hellmann 1985b).

As the amount of chlorinated organics adsorbed on the suspended solids fraction is very slight in comparison with the fraction in the water phase (~1%), analysis of suspended and settleable solids can as a rule be dispensed with. So long as there is no indication to the contrary, the same can apply to the sludges and bottom sediment.

Owing to the widespread use of chlorinated organic solvents, especially di- and trichloromethane and 1,1,2-trichlorotrifluoroethane, the pretreatment and working-up of the sample should as far as possible be performed in a completely solvent-free environment. The filling of the headspace sample containers and also the addition of reference materials should therefore be performed outside the laboratory 'in the open air'.

Frequently it is impossible to produce an organochlorine-free water for calibration purposes in the laboratory, as invariably minute amounts will be sucked in. A suitable alternative may be provided in the form of spring water. On the evidence of special investigations the sample may be taken by filling brown glass bottles at the sampling point and transported to the laboratory without any measurable change in the organochlorine concentrations. The samples can also be kept and stored for a few days in this state provided that the container has been absolutely filled and tightly closed.

A further problem area concerns the evaluation of the results of surface water analyses, as will be shown in the following section.

11.3 LOW-BOILING ORGANOCHLORINE COMPOUNDS IN SURFACE WATERS

The low-boiling chlorinated organic hydrocarbons in question are present almost exclusively in the water phase. This of course facilitates their analysis and the associated hydraulic evaluation. Thus one is not bothered by the variable suspended solids transport and its associated problems. Moreover sediment investigations are unimportant. Other types of complication do of course arise if one attempts to extrapolate an investigation initially confined to one sampling point to a downstream profile study. It involves the volatilisation of the organochlorine compounds, especially the more volatile components $CHCl_3$ and CCl_4, quite apart from CH_2Cl_2. This behaviour is closely related to the turbulence of the water and, less closely, to the water temperature. With these aspects in mind it will be apparent that the analytically determined flux of organochlorine compounds is by no means identical with the total amount discharged to the watercourse.

First of all let us look at the results for Koblenz. Figs 11.3 and 11.4 give a general

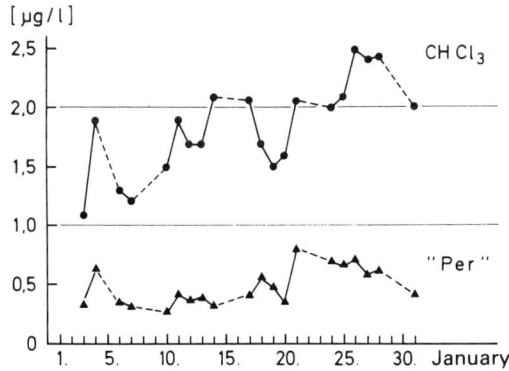

Fig. 11.3 — Concentrations of $CHCl_3$ and perchloroethylene (Per) in the Rhine at Koblenz during January 1983.

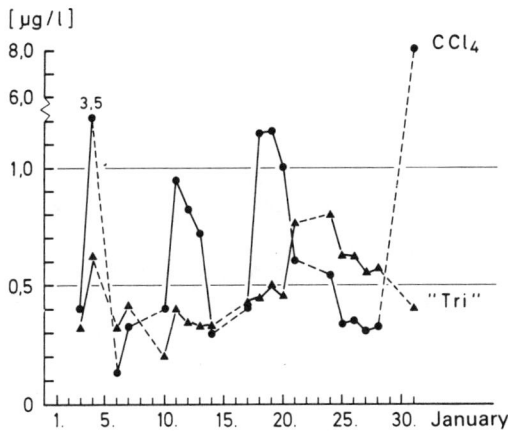

Fig. 11.4 — Concentrations of CCl_4 and trichloroethylene (Tri) in the Rhine at Koblenz during January 1983.

impression of the day-to-day concentration profiles and the variation encountered in the Rhine at Koblenz over a period of a month. In Fig. 11.3 the concentrations of perchloroethylene range from 0.3 μg/l to 0.8 μg/l with an average of 0.5 μg/l. A trend to higher or lower values is not apparent. The corresponding discharge values at around 2000 m^3/s were relatively constant, apart from a short-lived increase between 18 and 20 January. Chloroform behaved differently, inasmuch as it exhibited a rising trend from 1.0 to 2.5 μg/l. A point of particular note was the fall from 2.0 to 1.5 μg/l during the period of elevated discharge.

While trichloroethylene behaved similarly to perchloroethylene in respect of concentration levels (Fig. 11.4), for CCl$_4$ very pronounced fluctuations in individual values between 0.4 and 8 μg/l were recorded, which perhaps may be interpreted as evidence for a large discharge (the Lower Main) in the vicinity. April and May 1983 brought very high flows, which were largely reflected in the concentration pattern of chlorinated organics, and chloroform in particular—see Fig. 11.5.

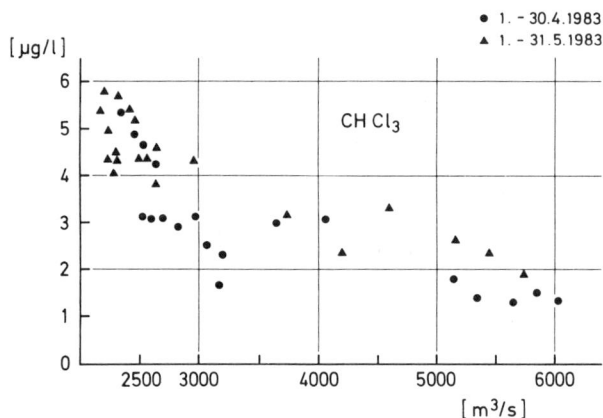

Fig. 11.5 — Relationship between CHCl$_3$ concentration and discharge in the Rhine at Koblenz, 1 April–31 May 1983.

The graphical relation between concentration and discharge such as indicated for chloroform in Fig. 11.5 frequently allows other important conclusions to be drawn. In this graph the organochlorine concentration declined with increasing discharge. Such a fall indicated a relatively uniform contamination of the water body, which under the given conditions must originate from a number of separate sources. The hyperbolic decrease of the CHCl$_3$ concentration implies an approximately constant mass flow rate. The three remaining compounds behave somewhat differently. Apart from CCl$_4$, the unsaturated solvents tri- and perchloroethylene apparently increase their mass flowrate as the discharge increases (Koblenz 1983), a subject which will not be discussed further here.

The repeated considerations of the relation between discharge and concentration permit the deduction that for exact mass transport estimates, the monthly averages are not always very effective. On the other hand monthly averages do provide an indication of the order of magnitude of the pollution level.

According to Fig. 11.6 this order of magnitude is about 2 μg/l for chloroform in

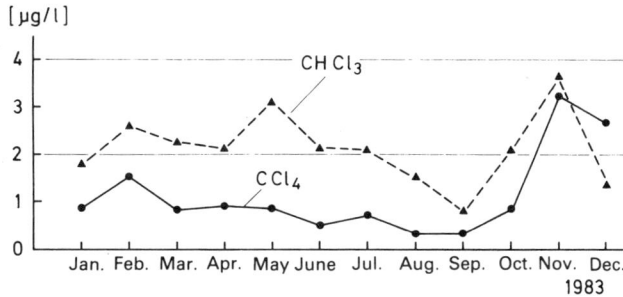

Fig. 11.6 — Mean monthly concentrations of $CHCl_3$ and CCl_4 in the Rhine at Koblenz, 1983.

the Rhine at Koblenz, and for CCl_4 mostly below 1 μg/l. Contrasting with pronounced minimum values for August/September (the holiday season?), there were maxima in November. For tri- and perchloroethylene the particular levels of contamination were between 0.5 and 1 μg/l, and below 0.5 μg/l, respectively (see Fig. 11.7).

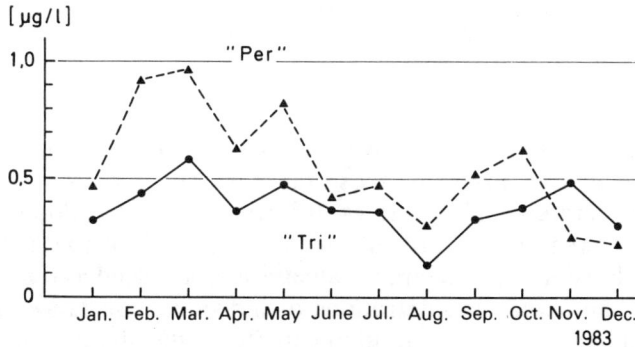

Fig. 11.7 — Mean monthly concentrations of Tri and Per in the Rhine at Koblenz, 1983.

Also in the case of these two solvents there was a distinct minimum with concentrations of 0.1–0.3 μg/l during August. The high levels for both compounds in the Rhine during February and March do not seem to fit into the picture, however. There may be a relation between concentration in surface waters and the winter season (increased consumption of cleaning agents? weather? temperature?). For mass transport calculations, the highest monthly value was obtained for May!

Where the mass fluxes are calculated from the monthly averages for discharge and concentration, then one should always be wary of the uncertainties connected with this method. As distinct from the correct, if tedious, process of daily mass transport estimates, in this case 'unreal' values may be obtained which may give rise to false interpretations. In Fig. 11.8, for example, one can observe an increase in the

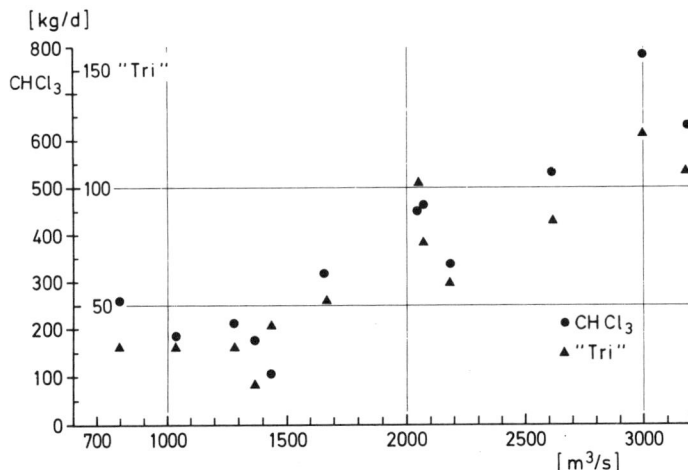

Fig. 11.8 — Relationship between daily transport rates and discharge for $CHCl_3$ and Tri in the Rhine at Koblenz in 1983, based on the monthly means.

$CHCl_3$ flux from 100–200 kg/d to over 500 kg/d based on the monthly average. For trichloroethylene the apparent mass transport rate increased from 50 to over 100 kg/d at moderately high flowrates. It is true that for trichloroethylene the mass transport rate increased with discharge in the Rhine at Koblenz during 1983. In Table 11.2 the calculated monthly average values are reproduced as *rates per second.*

Although the actual concentrations of individual substances in the Rhine at Koblenz exhibited a very low variation amplitude, this should not deceive one into making false conclusions. For on the strength of longitudinal profile studies the concentration of organochlorines frequently appears to vary widely. Thus localised peak values rapidly decline to a low level as one moves downstream, and certainly not principally because of dilution and mixing of effluent plumes. Figs 11.9 and 11.10 become intelligible only if following the input of considerable quantities of organo-chlorines their elimination then occurs by escape to the atmosphere. In contrast to the high-boiling materials (see Fig. 6.11), therefore, the organochlorine concent-ration does not necessarily increase in the direction of flow. High organochlorine concentrations in the lower reaches or outflow region of a river may, of course, be

Table 11.2 — Organochlorine mass transport rates in the Rhine at Koblenz 1983
(monthly averages)

1983 Calendar month	Discharge (m^3/s)	Water temp. (°C)	Mass flux (g/s)			
			$CHCl_3$	CCl_4	Tri	Per
Jan.	2190	7.0	3.96	2.04	0.70	1.05
Feb.	2070	5.5	5.32	3.15	0.91	1.90
Mar.	2050	8.2	5.22	1.54	1.21	2.00
Apr.	3440	11.5	7.57	2.79	1.24	2.20
May	3000	14.3	9.60	2.34	1.44	2.52
Jun.	2610	19.5	6.10	1.25	1.00	1.10
Jul.	1670	23.0	3.70	1.20	0.60	0.80
Aug.	1310	23.0	2.03	0.44	0.19	0.41
Sep.	1440	19.0	1.21	0.49	0.48	0.75
Oct.	1040	16.0	2.18	0.89	0.40	0.65
Nov.	795	9.8	2.96	2.58	0.39	0.20
Dec.	1280	6.3	2.44	3.43	0.38	0.28

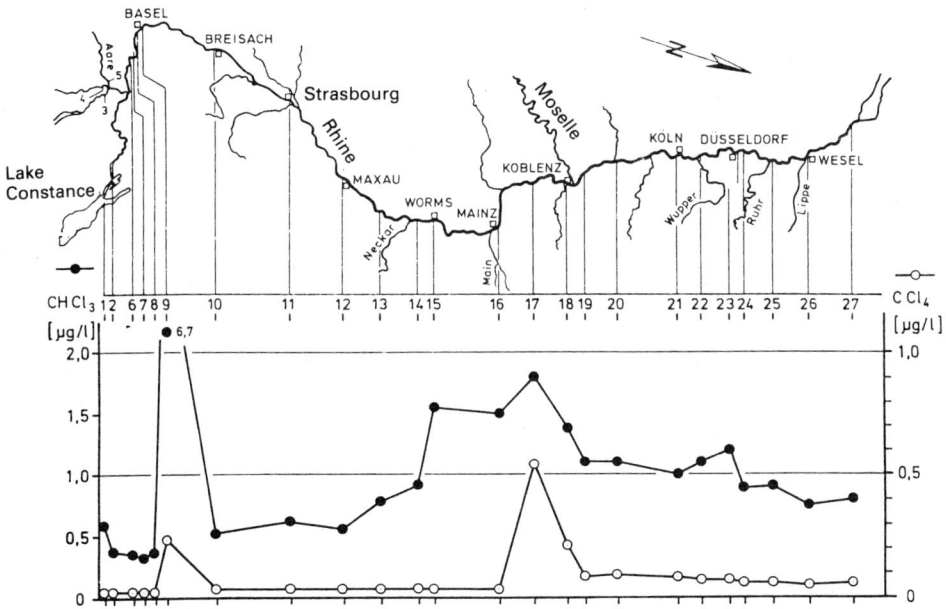

Fig. 11.9 — Longitudinal profile investigations of $CHCl_3$ and CCl_4 in the Rhine, 9–13 April 1984.

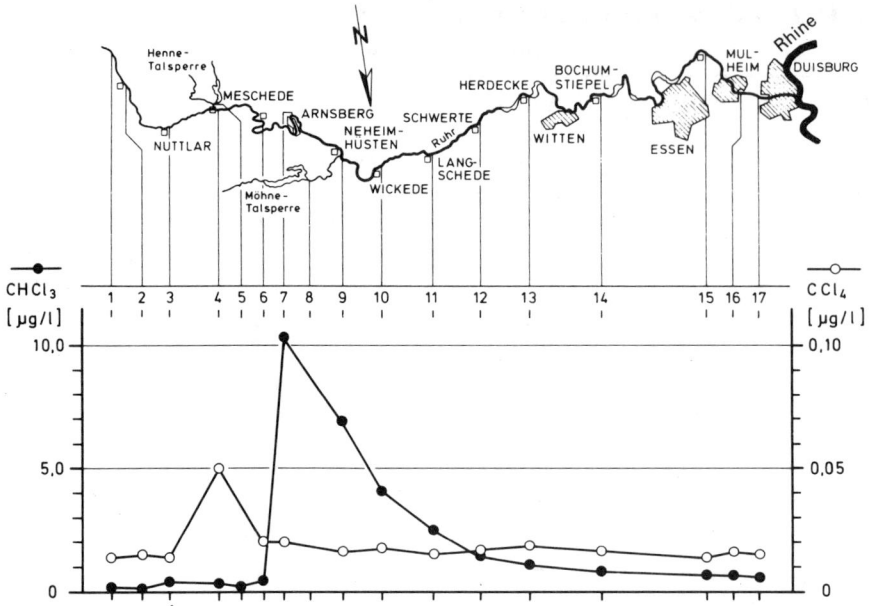

Fig. 11.10 — Longitudinal profile study of CHCl$_3$ and CCl$_4$ in the Ruhr, 12–13 July 1984.

expected where a high population density or, more especially, industrial and trade establishments render the discharge of effluents in this region inevitable (Fig. 11.11).

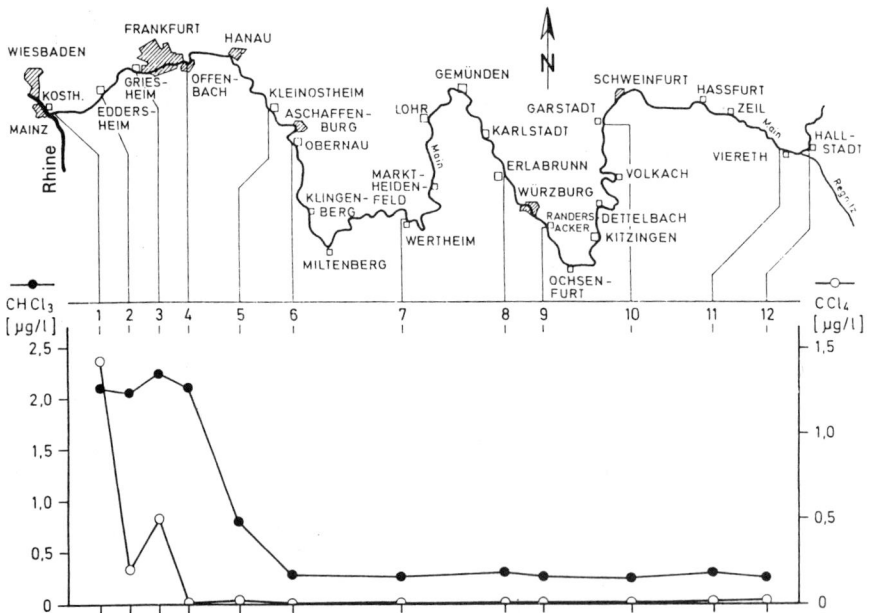

Fig. 11.11 — Longitudinal profile study of CHCl$_3$ and CCl$_4$ in the Main, 27–28 March 1984.

LITERATURE

11.1 and 11.2

Malle, K.-G. (1984). Die Bedeutung der 129 Stoffe der EG-Liste für den Gewässerschutz, Z. Wasser-Abwasser-Forsch. **17**, 75–81.

Stieglitz, L., Roth, W. (1976). Das Verhalten von Organohalogenverbindungen bei der Trinkwasseraufbereitung, Vom Wasser **47**, 347–377.

Bätjer, K. u. a. (1980). Analyse und Verteiling von leichtflüchtigen halogenierten Kohlenwasserstoffen in Bremmer Trinkwasser,Vom Wasser **54**, 143–159.

Dietz, F., Traud, J. (1973). Bestimmung niedermolekularer Chlorkohlenwasserstoffe in Wässern und Schlämmen mittels Gaschromatographie, Vom Wasser **41**, 137–155.

Ministerium für Ernährung, Landwirtschaft, Umwelt und Forsten Baden-Württemberg (1983). Leitfaden für die Beurteilung und Behandlung von Grundwasserverunreinigung durch leichtflüchtige Chlorkohlenwasserstoffe, Stuttgart, S. 65–89, dort weitere Literatur-Hinweise.

Trenel, J., Wolf, G., Neumayer, V. (1983?). Organohalogenverbindungen im Wasserkreislauf – Analysenverfahren, Probenahmen und Bewertungen im Abwasserbereich und im Vorfluter. Bericht für das Bundesinnenministerium, Institut für Wasser-, Boden- und Lufthygiene des Bundesgesundheitsamtes, Berlin.

Dietz, F., Traud, J., Koppe, P. (1982). Leichtflüchtige Halogenkohlenwasserstoffe in Abwässern und Schlämmen, Vom Wasser **58**, 187–205.

Zürcher, F., Giger, W. (1976). Flüchtige organische Spurenkomponenten in der Glatt, Vom Wasser **47**, 37–55.

DIN 38407 Teil 4 (in Vorbereitung–1985). Gemeinsam erfaßbare Stoffgruppen (Gruppe F); Bestimmung von leichtflüchtigen Halogenkohlenwasserstoffen in Wasserproben (F4).

Rook, J. J., Meijers, A. P. u. a. (1975). Head-space-Analyse flüchtiger Spurensubstanzen im Rhein, Vom Wasser **44**, 23–30.

Dampfraum-Analysator F 45: Druckschrift der Firma Perkin-Elmer Nr. 1808/9.78 Überlingen/Bodensee.

Kolb, B., Auer, M., Pospisil, P. (Mai 1982). Quantitative head-space-Analyse flüchtiger Halogenkohlenwasserstoffe (HKW) aus wässrigen Proben, Druckschrift Bodenseewerk Perkin-Elmer & Co. GmbH, Überlingen/Bodensee.

Hellmann, H. (1985a). Anwendung des "Head-Space"-Verfahrens auf die Bestimmung leichtflüchtiger chlorierter Kohlenwasserstoffe in Oberflächengewässern, Z. Wasser-Abwasser-Forsch. **28**, 92–98.

Grob, K., Zürcher, F. (1976). Stripping of trace organic substances from ater. Equipment and procedure, J. Chromatogr. **117**, 285–294.

Schulz, J. (1984). Gaschromatographische Bestimmung leichtflüchtiger Halogenkohlenwasserstoffe (LHKW) im Trinkwasser, GIT-Supplement Umweltschutz-Umweltanalytik **1**, 50–52.

Hellmann, H. (1985b). Zur Adsorption und Desorption leichtflüchtiger Chlorkohlenwasserstoffe am Beispiel von Sedimenten und Schwebstoffen, Dtsch. Gewässerkd. Mitt. **29**, 111–115.

11.3

Battelle-Institut e.V. Frankfurt am Main (1977). Mainwasseruntersuchungen, Bericht zum Auftrag 198–13, Frankfurt.

Kußmaul, H., Mühlhausen, D. (1980). Organohalogenverbindungen im Untermain, WaBoLu-Ber. **7**, 19–22.

Bundesanstalt für Gewässerkunde (1984). Auftreten, Herkunft und Verhalten von leichtflüchtigen Chlorkohlenwasserstoffen in Oberflächengewässern der Bundesrepublik Deutshland, BfG-0159, Koblenz.

Hellmann, H. (1984). Leichtflüchtige Chlorkohlenwasserstoffe in den Gewässern der Bundesrepublik Deutschland – Auftreten und Bilanz, Haustechnik-Bauphysik-Umweltschutz-Gesundheitsingenieur **105**, 269–278.

Hellmann, H. (1985). Verhalten von leichtflüchtigen Chlorkohlenwasserstoffen in Oberflächengewässern der Bundesrepublik Deutschland, Z. Wasser-Abwasser-Forsch. **18**, 210–216.

Bauer, U. (1976). Flüchtige organische Chlorverbindungen im Trink- und Oberflächenwasser. Ergebnis von Summenbestimmungen. Vortrag bei der 6. Arbeitstagung der Deutschen Gesellschaft Für Hygiene und Mikrobiologie in Mainz am 28.9.76.

Arbeitsgemeinschaft für die Reinhaltung der Elbe (1983). Chlorierte Kohlenwasserstoffe – Daten der Elbe – 1980–1982, Hamburg.

DVGW (1981). Organische Schadstoffe in den Fließgewässern der Bundesrepublik Deutschland, Forschungsvorhaben Wasser IIA 77 für das Bundesinnenministerium Frankfurt am Main.

Kußmaul, H., Fritschi, U., Fritschi, G., Schinz, V. (1978). Leichtflüchtige Halogenkohlenwasserstoffe im Rheinwasser, Uferfiltrat und Trinkwasser, WaBoLu-Ber. **3**, 75–85.

IAWR Bericht der Jahrestagung 1979 und ARW Jahresbericht 1976.

Kühn, W., Sander, R. (1978). Vorkommen und Bestimmung leichtflüchtiger Chlorkohlenwasserstoffe, Hydrochem. hydrogeol. Mitt. **3**, 327–340.

12

Detergents

12.1 PRELIMINARY REMARKS

During 1983, according to data provided by the Manufacturers' Association a grand total of 350 000 tonnes of detergents was produced in West Germany. This figure can be broken down into the respective totals for anionic detergents (about 160 000 t), nonionic detergents (115 000 t) and cationic detergents (30 000 t). Among the anionics the predominant types were the alkylbenzenes and alkanesulphonates, and for nonionics both alcohol and alkylphenol ethoxylates. The group of cationic detergents constitutes a special case, as it is not concerned with surface-active materials in the strict sense, and comprises almost entirely (85–90%) the single compound distearyldimethyl-ammonium chloride (DSDMAC).

Analytically the anionic detergents are as a rule estimated, using the method of DIN 38 409 Part 23, as methylene-blue active substances (MBAS) and the nonionics as bismuth-active substances (BiAS) according to the same standard. These two standard methods are incorporated in the German Legal Ordinance regarding household detergents. For cationic detergents on the basis of the ammonium ion, the disulphine blue method (Waters and Kupfer, 1976) is available.

In the context of household detergent analyses and including the testing of biochemical degradability by means of the OECD Confirmatory Test, the stated methods have proved entirely satisfactory. For the analysis of detergents in waste-waters, and more particularly in surface waters, it appears desirable to strip the detergents from the aqueous phase (Wickbold, 1971). By this means, on the one hand larger volumes can be analysed, and on the other undesirable interfering compounds can be eliminated or at least suppressed. The detergents are taken up in acetic ester and after the transfer worked up according to the prescribed methods.

As has been discovered, air-stripping does not lead to a completely pure detergent fraction, so that further separation pretreatment must be performed. Further problems arise in connection with detergent estimation in sludges, not merely the bacterial sludge from Confirmatory Test equipment but also the heterogeneous biomass obtained from municipal sewage treatment installations. Such problems can only be solved by means of suitable and effective pretreatment.

12.2 DETERGENTS AND BIOGENIC INTERFERING SUBSTANCES

Analytical results for detergents in Rhine water between 1971 and 1977, obtained without the air stripping step of the DIN prescribed method, are summarised in Table 12.1. On examination, these indicate a clear trend from higher to lower values,

Table 12.1 — Frequency distribution of MBAS concentrations in the Rhine at Koblenz

Calendar year	MBAS (mg/l) ranges					
	<0.10	≥0.10	≥0.15	≥0.20	≥0.30	≥0.40
	numbers of samples					
1971	0	26	20	16	9	2
1972	2	24	18	12	4	0
1973	1	25	16	11	1	1
1974	3	23	15	8	2	2
1975	15	11	5	1	0	0
1976	11	15	3	0	0	0
1977	19	7	1	0	0	0

Note: Columns 1 and 2 are additive; $n=26$. Remaining columns are cumulative from Column 6 to Column 2.

from year to year. For example, in 1971 26 spot samples were taken at fortnightly sampling intervals, of which 20 showed MBAS contents greater than 0.15 mg/l; in 1977 only one sample exceeded this value. Similarly in the Rhine at the Dutch frontier, the concentrations also decreased steadily between 1974 and 1980; from 1980 up to 1984 the concentration remained more or less constant (Fig. 12.1).

Of course such deductions are risky without any reference to the discharge (cf. Section 1.1) and thus only tentative. On the other hand, calculations of MBAS transport rates, in which the discharge data were incorporated, produced the same conclusion (Fig. 12.2). That the decline in the MBAS contamination in the Rhine (and in other German rivers) is due to a decrease in the actual quantities of anionic detergents carried by the river has also been reported by Fischer (1980).

The residual MBAS concentrations in the Middle and Lower Rhine from 1978 to 1980 appear, following more exact hydraulic investigations, to be independent of the discharge. It does not diminish with increasing discharge, as would be expected to occur from the input of anthropogenic substances (see Section 6.4). No doubt this type of behaviour in the MBAS values is indicative of the effect of interfering substances on the results. A further indication can be obtained from a study of the longitudinal profiles.

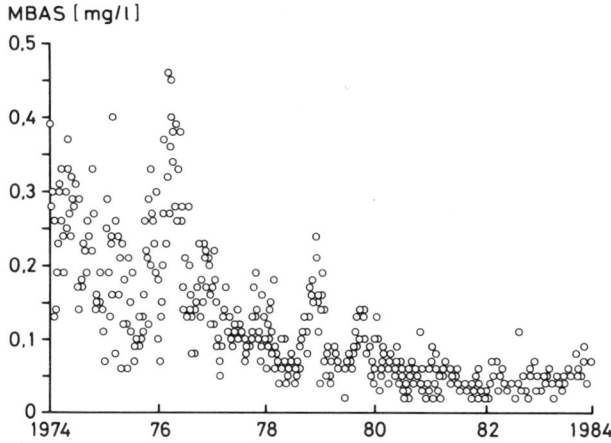

Fig. 12.1 — MBAS contents in the Rhine at Bummen/Lobith for 1974–1984 (from Table of Numerical Data).

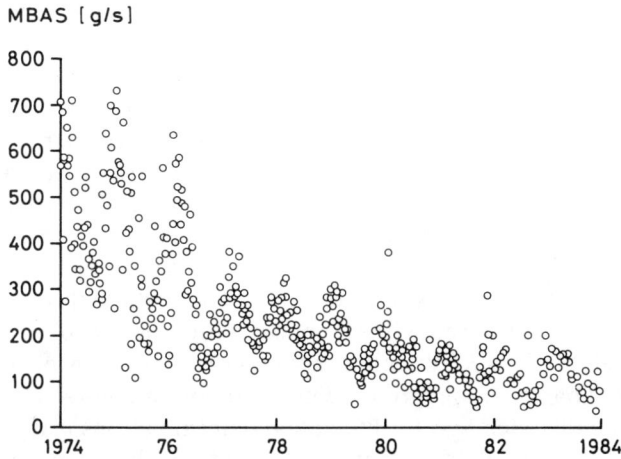

Fig. 12.2 — MBAS transport rates in the Rhine at Bummen/Lobith during 1974–1984 (from Tables of Numerical Data).

Besides the 'classical' case of the concentration increase for nitrates, for example (Fig. 6.11), and the special case of the low-boiling chlorinated hydrocarbons (Fig. 11.9), we have a further variant in this case. Thus one often obtains roughly equal MBAS concentrations, irrespective of the sampling point. Occasionally one hears the remark in specialist circles that this finding can be explained on the basis of biochemical decomposition. According to this hypothesis, the detergents introduced to the river at a given point are almost completely degraded before the next point source discharge is reached.

While one must not underestimate the biochemical decomposition of organic matter in surface waters, particularly at summer temperatures (Fig. 8.9), nevertheless it is a fact that the presence of biogenic MBAS in roughly the same order of magnitude inhibits the making of reliable deductions concerning detergent pollution in a longitudinal profile, if the analyses are performed by the standard method.

Regarding the nonionic detergents, far too few results are available in the literature. However, there is no doubt that, just as for the anionic detergents, interfering substances are included in the results of measurements by the standard method of determination. Only in this way is it possible to understand the strong increase in the load of bismuth-active substances (BiAS) shown later in this chapter in Table 12.3.

So far as can be predicted, for the near future the exact determination of detergents in surface waters in the lower microgram/litre range is unlikely to become a routine laboratory task.

12.3 ANALYSIS OF ANIONIC DETERGENTS

Departing slightly from the general objective of this monograph, the state of the art for analysis of anionic detergents, as far as the basics are concerned, will now be discussed. The analytical method is based on the formation of a complex with methylene blue, disulphine blue, or the so-called Dragendorff reagent, although the structures of the complexes do not appear to be understood. The chief problem in the realm of detergent analyses in natural waters lies at present not so much in the meaningful interpretation of the data, nor in the process of sampling, but in the method of obtaining pertinent analytical results. As long as the detergent concentration is of the order of 0.3–0.5 mg/l and over, it is sufficient to deduct a constant amount of, e.g., 0.05 mg/l from the result to allow for the 'imaginary' MBAS content (Fischer) or the biogenic background value. Understandably this is no longer applicable when the measured MBAS concentration lies within the region of this background level. Under such circumstances there are two conceivable alternatives. Thus one may abandon the group determination methods for the respective detergent types and instead go for individual determinations using GC, GC/MS or HPLC techniques. Actually the efforts at present are chiefly in this direction. The other possibility involves a chromatographic separation of the MBAS (or BiAS) compounds into a relatively pure detergent fraction, which then can be analysed, together with a biogenic fraction, for which the chemical composition and structural elucidation constitute a special task for qualified laboratories.

The second approach seems at first to be simpler and more promising. It is, however, a valid question whether one proceeds via MBAS complex formation or dispenses with the use of methylene blue altogether. At first, attempts were made to separate the blue extract on silica gel plates (F-60) with various solvents, hoping that a pure detergent fraction would result. According to Fig. 12.3 the extract can be resolved with the aid of $CHCl_3/CH_3OH$ (9:1, v/v) into several coloured zones, depending on its source or composition, which could then be evaluated using a scanner connected to a printer to give the corresponding chromatogram. For the upper chromatogram of Fig. 12.3, 3 mg of 'Pril' [a proprietary blend ?–Ed] were reacted with methylene blue, while the middle chromatogram indicates the MBAS from 5 l of spring water, and the lowest one shows the corresponding pattern for 2 l of water from the Rhine at Koblenz in 1977.

When the positions of the various coloured zones are compared, the following two points emerge. The anionics, where they occur, remain in the vicinity of the point of application. However, biogenic MBAS substances are also located in this area (middle chromatogram), even though a number of coloured interfering reaction products have already been separated. All attempts using other solvents to obtain a pure detergent–methylene blue fraction on the adsorbent plate have met with failure. Thus there arose first of all the necessity for clearly identifying the anionics and intruding substances as part of the separating process, and then to optimise the analytical method, on this basis. In anticipation of the preliminary outcome it must be remarked that a separation of the blue complex in the desired manner was unsuccessful. Nevertheless the reaction with methylene blue may be useful, e.g. as a first step in the enrichment process when working up the sample.

For purposes of identification, IR-spectrometry presented itself. As the methylene blue cation interferes with strong bands in the so-called fingerprint region, this had to be removed and replaced by less disrupting ions such as H^+ or NH_4^+. One can in fact separate the methylene blue by cation exchange directly on the silica gel plate (Hellmann 1983). For this purpose the MBAS extract applied by streaking onto the

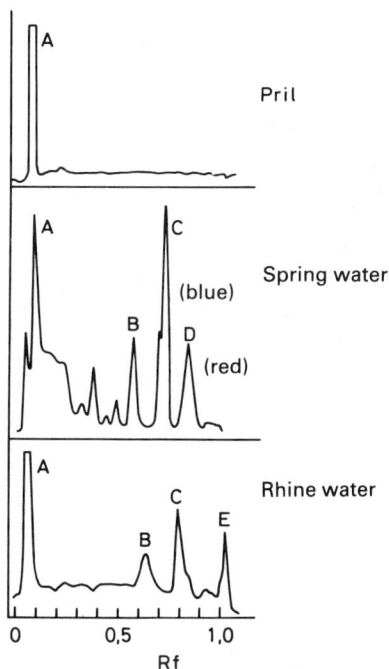

Fig. 12.3 — Chromatography of MB extracts on silica gel F-60. Picture showing absorption in remission at 665 nm with a chromatograph spectrophotometer.

adsorbent plate is developed upwards by $CH_3OH/2N$ NH_4OH (3:1) so that the detergent anions migrate in the direction of the solvent flow, probably in the form of NH_4^+. The fractions situated at the solvent front in most cases contain further unwanted substances which entered the extract during extraction of the sample with $CHCl_3$, such as fatty acid esters, alkanes and suchlike. These can also be separated on the plate, for which the solvents $CHCl_3$ or $CHCl_3/CH_3OH$ (9:1) may be employed. The interfering substances migrate upwards and leave the MBAS (although without the original MB) behind. One then removes the MBAS zone from the glass support, pulverises it and feeds the powder into a small glass column, which is then eluted with $CHCl_3/CH_3OH$ (1:1). After taking up in $CHCl_3$ the IR spectrum can be obtained.

The process sequence alluded to is summarised as a flow chart in Fig. 12.4. The mixture of methylene blue-active substances isolated by the process described will contain the desired anionic detergent in addition to the associated biogenic compounds, for which the IR spectrum provides further information.

Anionic detergents of the alkylbenzene-sulphonate (ABS) type give a spectrum of the kind shown in Fig. 12.5, while biogenic MBAS give spectra of the type of Fig. 12.6.

While in the larger surface water systems the biogenic MBAS fraction predominates, and anionic detergents can thus be isolated in the pure state only with difficulty — a situation which applies also to sediments, river sludge deposits and suspended solids — the quantitative determination of anionics in sewage sludges as a rule

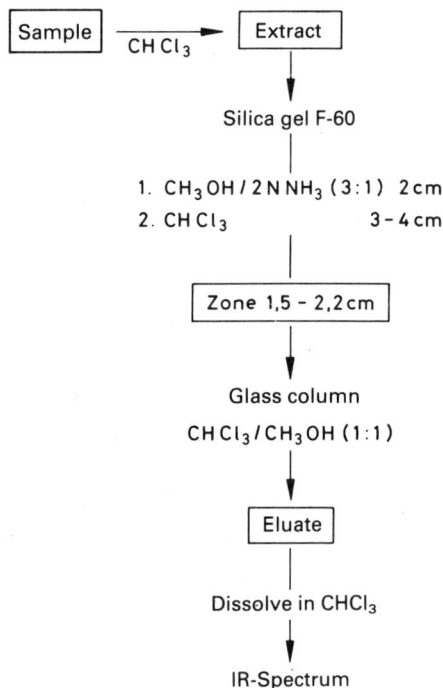

Sample — $CHCl_3$ → Extract

↓

Silica gel F-60

1. $CH_3OH / 2N\,NH_3$ (3:1) 2 cm
2. $CHCl_3$ 3 – 4 cm

↓

Zone 1,5 – 2,2 cm

↓

Glass column
$CHCl_3 / CH_3OH$ (1:1)

↓

Eluate

↓

Dissolve in $CHCl_3$

↓

IR-Spectrum

Fig. 12.4 — Flow chart for isolation of MBAS preparatory to spectroscopic identification.

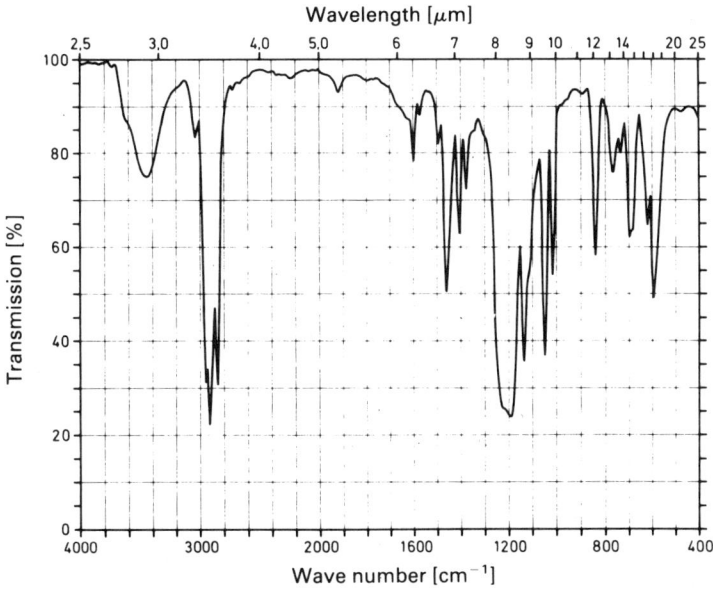

Fig. 12.5 — IR spectrum of Alkylbenzene sulphonate in sediment sample.

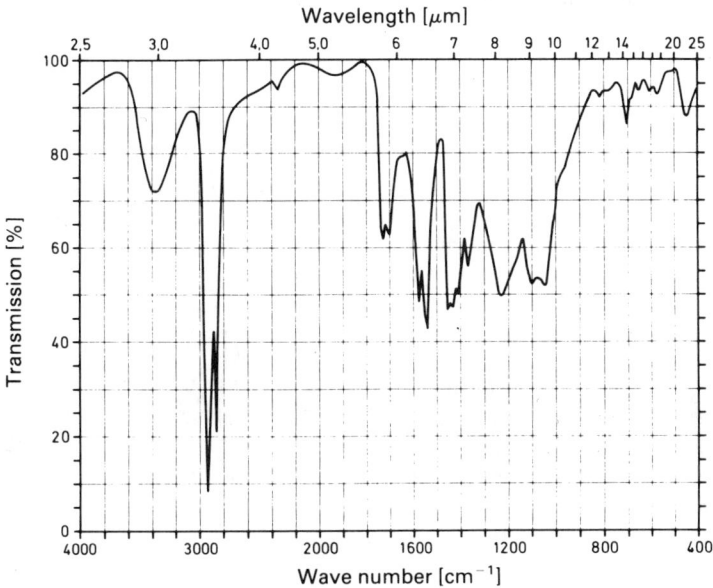

Fig. 12.6 — IR spectrum of biogenic MBAS in a sludge sample (cf. 12.5).

presents fewer difficulties. In any case, however, the biogenic fraction must be separated from the anionic detergent fraction where the IR spectrum indicates the presence of both groups. For this purpose one subjects the mixture obtained by the process of Fig. 12.4 to renewed fractionation on silica gel and proceeds further according to the second alternative path, described more fully below.

Experience also shows that, of the group of anionic detergents, the alkylbenzene sulphonates are practically the sole members to occur in surface waters, at least in the form of 'intact' molecules. The other materials such as alkane sulphonates and sulphates which are used in household detergents appear to be much more rapidly degradable.

As previously mentioned, alternatives to the reported method of chromatography of methylene blue extracts are applicable. Thus the acetic ester extract obtained following air-stripping, or a methanol extract of sludge or sediments, after transfer to CHCl$_3$ (which is generally advisable) can be applied to the silica gel layer without the formation of a complex with methylene blue. Under these conditions the ion-exchange step is eliminated. Development takes place either with CHCl$_3$/CH$_3$OH (6:1) over 10 cm, or with CHCl$_3$/CH$_3$OH (3:1) over 5 cm. The precise position of the anionic detergent fraction on the thin layer depends on a number of parameters and cannot be regarded as constant. Thus following the application of the extract to the adsorbent layer, a scanner is used to ensure that besides the line of application a particular point is found which after development and spraying of this region with Rhodamine B (0.03% in water) or Pinacryptol-yellow will indicate the position of the anionic detergent. A reliance on reference substances can lead to errors as the detergent molecules present in surface waters or sludges may not tally exactly, with respect to composition and migration distance on the silica gel plates, with the original constituents or reference materials.

In addition, in impure detergent fractions (Fig. 12.7) alkylbenzene sulphonates

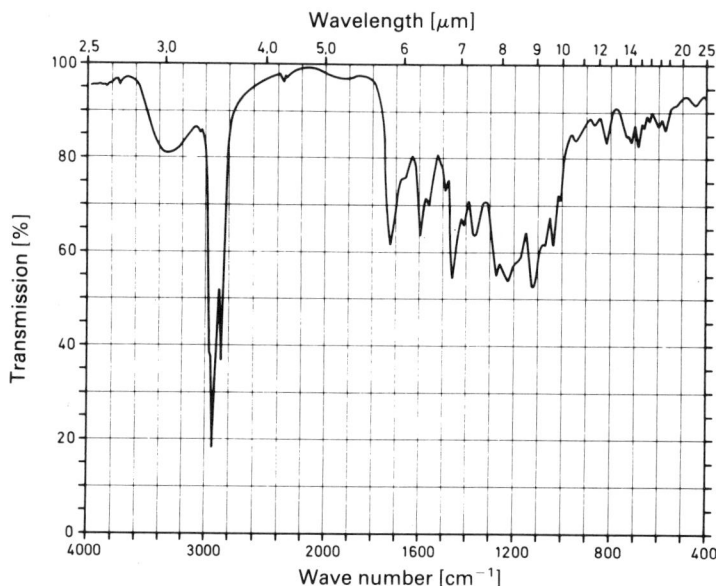

Fig. 12.7 — Nearly pure anionic detergent fraction from a sample of water from the Saar and its tributaries.

may be recognised by the key absorption bands at 1603, 1200, 1045, 1003, and 832 cm^{-1}. A necessary condition is a certain minimum concentration.

Small rivers may from time to time exhibit higher MBAS concentration values-(Table 12.2). At levels of over 0.2 mg/l the presence of anionics can be inferred with

Table 12.2 — MBAS in the Saar and tributaries, 4–8 December 1984

Sampling point	MBAS	o-PO$_4^{3-}$ (as P, mg/l)	total PO$_4^{3-}$ (as P, mg/l)
Saar			
Güdingen	0.12	0.43	0.48
Luisenthal	0.80	0.50	0.68
Völkingen	0.38	0.58	0.72
Hostenbach	0.35	0.65	0.80
Bous	0.41	0.73	0.93
Lisdorf	0.32	0.71	1.80
Dillingen	0.31	0.70	0.96
Fremersdorf	0.20	0.43	0.58
Mettlach	0.25	0.49	0.71
Serrig	0.20	0.47	0.67
Schoden	0.15	0.44	0.61
Kanzem	0.14	0.46	0.63
Tributaries			
Blies	0.14	0.50	0.62
Saarbach	0.26	0.75	0.92
Rohrbach	1.39	1.85	3.57
Sulzbach	2.84	5.33	6.40
Fischbach	0.60	0.76	1.13
Köllerbach	0.73	2.00	2.25
Rossel	0.74	2.20	2.80
Bist	0.36	0.85	1.36
Prims	0.13	0.19	0.38
Nied	0.10	0.26	0.32
Seffersbach	0.09	0.16	0.31
Leuk	0.10	0.19	0.27

some certainty. (Indications of their origin in sewage effluent discharges may be given by simultaneously increased levels of PO3−4.) If one treats the samples referred to in Table 12.2 according to the method outlined in Fig. 12.4 and also to the direction in the text, then one obtains a spectrum similar to Fig. 12.7, in which anionics can be detected. After fractionating further, the anionics in the form of alkylbenzene sulphonates are clearly recognisable (Fig. 12.8).

Finally, in addition to spectrophotometeric methods, IR spectrometric anionic detergent determination can be performed with reference to the CH_2 vibration band at 2925/2928 cm^{-1}. The sensitivity of detection by this method is, however, somewhat inferior. It is also important for the analyst to realise that anionic detergents in surface waters, just like nonionics, are transported almost entirely in the water phase. In sediments and sludge samples, therefore, these detergents may frequently not be detected.

12.4 ANALYSIS OF NONIONIC DETERGENTS

With respect to the amounts produced and used in the German Federal Republic, the nonionics are not far behind the anionic detergents; they are therefore of relevance to surface water analysis. If one subjects the DIN standard process to intensive investigation, then like the MBAS method for anionics, it proves to be *non-specific* for detergents. The process of precipitation with the Dragendorff reagent in its modified composition, $KBa[BiI_4]_2$, applies exclusively to ethylene oxide or alkylene oxide derivatives and hence to compounds incorporating the C–O–C group which, as will be seen, is widely distributed in nature. Interference with the detergent estimation due to such compounds occurs in connection with the analysis of water, sludge and sediment samples. Water samples are ideally treated according to the scheme outlined in Fig. 12.9. The ultimate determination of bismuth, which according to the DIN method is carried out potentiometrically, may just as well be carried out by AAS or X-ray fluorescence methods, assuming a pure detergent fraction. Unfortunately in surface waters, as elsewhere, the same difficulties arise due to biogenic BiAS as are encountered with biogenic MBAS for anionic detergents (Hellmann 1985). This is disclosed by the oft-mentioned hydraulic evaluation (Table 12.3).

From the figures in Table 12.3 the BiAS concentration values for the Rhine at Koblenz in 1983/84 appeared practically independent of the discharge, although the scatter of the values in the range 0.014–0.070 mg/l was relatively large. A distinguishing feature, which suggests the presence of a biogenic component, is the increase of the BiAS flux with discharge. For verification of such a hydraulically based prediction, some confirmatory analytical evidence is required. Once again this is furnished by IR spectrometry. As the Dragendorff precipitate is soluble in $CH_3OH/2N$ NH_4OH (3:1, v/v) the total spectrum can be obtained, after taking up into $CHCl_3$, on a KBr disc. Even though the precipitation of the BiAS can be compared with the complex formation of anionics with methylene blue, nevertheless there are in this case some advantages for the analyst. Thus the $Ba[BiI_4]_2$ complex ion does not interfere in the IR region and hence does not have to be removed beforehand. On the other hand, the precipitate, after the prescribed wash with acetic acid, is largely free of impurities, including anionic detergents. The residual interference must therefore be attributable to the co-precipitated biogenic BiAS materials. Where cationic detergents are also present in noticeable amounts, these will be present as a Dragendorff complex in the precipitate. The obvious question then is: what do the IR spectra of the nonionics and the interfering substances look like and how can a separation from the cationics and the biogenic materials from the nonionics be achieved?

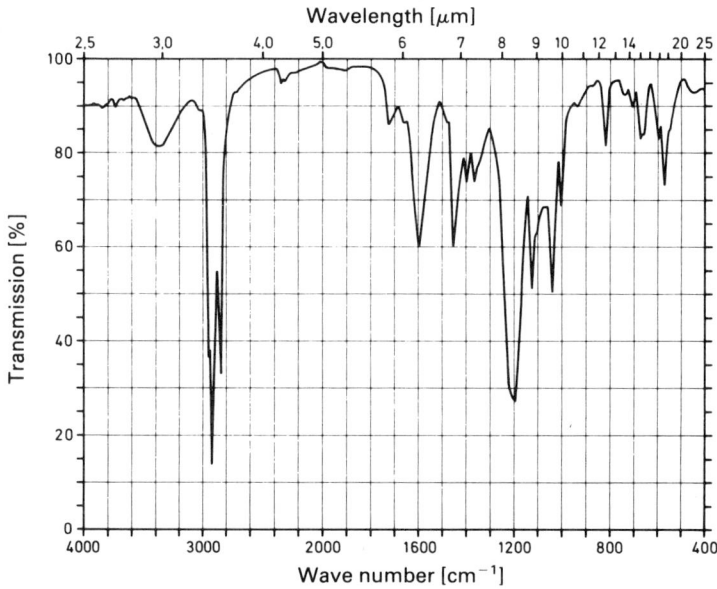

Fig. 12.8 — Anionic detegents from the Saar and tributaries, December 1984.

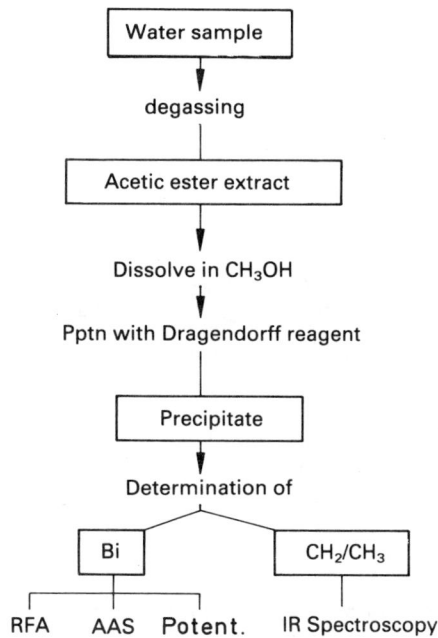

Fig. 12.9 — Flow chart for the determinatino of BiAS in water samples.

Table 12.3 — BiAS contents and transport rates in the Rhine at Koblenz

Date	Discharge (m^3/s)	BiAS $(\mu g/l)$	BiAS (g/s)
1988			
23 Nov.	620	50	31.0
29 Nov.	1500	28	42.0
5 Dec.	1450	60	87.0
6 Dec.	1210	30	36.3
13 Dec.	960	42	40.3
1984			
5 Jan.	1360	50	68.0
9 Jan.	1290	62	80.0
13 Jan.	1200	70	84.0
16 Jan.	2400	30	72.0
19 Jan.	2900	29	84.1
2 Feb.	1920	45	86.4
6 Feb.	2700	25	67.5
13 Feb.	3100	46	143
24 Feb.	1410	40	56.4
27 Feb.	1320	43	56.8
28 Feb.	1310	40	52.4
8 Mar.	1160	35	40.6
12 Mar.	1120	44	49.3
27 Mar.	1260	44	55.4
9 Apr.	1900	28	53.2
24 Apr.	1390	27	37.5
7 May	1600	14	22.4
21 May	1550	26	40.3
4 Jun.	2630	16	42.0
18 Jun.	1750	25	43.8
2 Jul.	2000	25	50.0
16 Jul.	1690	16	27.0
30 Jul.	1480	26	38.4
13 Aug.	1910	28	53.5
27 Aug.	1310	25	32.8
10 Sep.	1920	31	60.0
24 Sep.	2250	30	67.5
8 Oct.	—	46	—

According to information from the detergent manufacturers, in 1981 55 000 t of alcohol ethoxylates (EO) and 20 000 t of alkyl phenol ethoxylates were consumed in the German Federal Republic. In addition there were about 6000 t of EO/PO adducts:

$$-(EO)_n = (-CH_2-CH_2-O-)_n$$
$$-(PO)_n = (-CH-CH_2-O-)_n$$
$$\qquad\qquad\quad |$$
$$\qquad\qquad\ CH_3$$

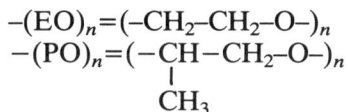

which on account of their poor degradability come under the 'transition order', and 20 000 t of nonionic detergents of various chemical groups.

In the following we will confine ourselves to the alcohol and alkylphenol ethoxylates. As a preliminary remark, the terms 'oxyethylate' or 'polyglycol ether' are sometimes employed instead of ethoxylate in the relevant literature.

The IR spectrum of the nonylphenol ethoxylate NP(EO)$_9$ is shown in Fig. 12.10

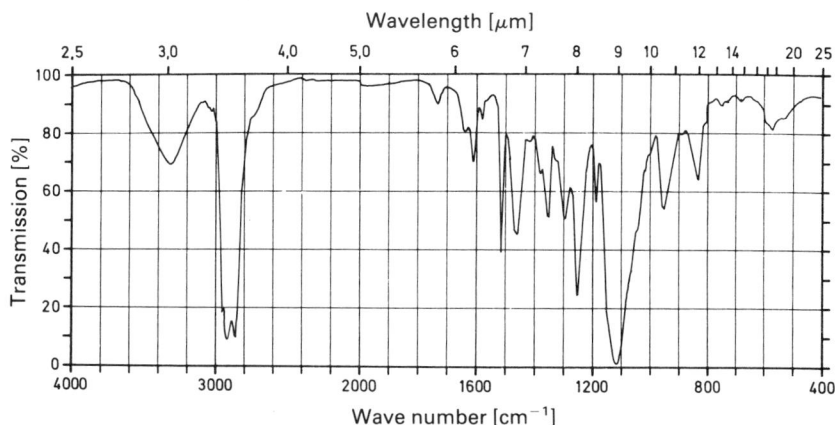

Fig. 12.10 — Scanning spectrum of nonylphenolethoxylate after precipitation with Dragendorff reagent on KBr.

and is characterised chiefly by the dominant band around 1101 cm^{-1} which represents the vibration of the aliphatic ether linkage C–O–C. Furthermore the aromatic bands 1609 and 1512 cm^{-1} are prominent. For the linear chain structure of the alcohol ethoxylates these bands are absent (Fig. 12.11) while the C–O–C band at 1100 cm^{-1} appears even more pronounced. Besides the similarly heavy CH$_2$/CH$_3$ bond vibration bands between 2800 and 3000 cm^{-1}, there are weaker bands at 950 and 850 cm^{-1}. Bismuth-active substances in detergent-free waters yield IR spectra like that in Fig. 12.12. The predominant ether band in this case is at 1094 cm^{-1}. In addition there is a carbonyl band (1736 cm^{-1}) as well as the so-called long-chain paraffin band (721 cm^{-1}). Up to but excluding the carbonyl band the spectrum resembles that of the alcohol ethoxylates.

In surface waters, BiAS of natural origin is always characterised by an IR spectrum of the type shown in Fig. 12.12, in which the ratio of the extinctions due to the C–O–C, the C=O and the CH$_2$/CH$_3$ bands can vary. Hitherto it has not been possible to distinguish alcohol ethoxylates from biogenic BiAS, and experiments involving thin-layer chromatographic separation have so far proved unsuccessful.

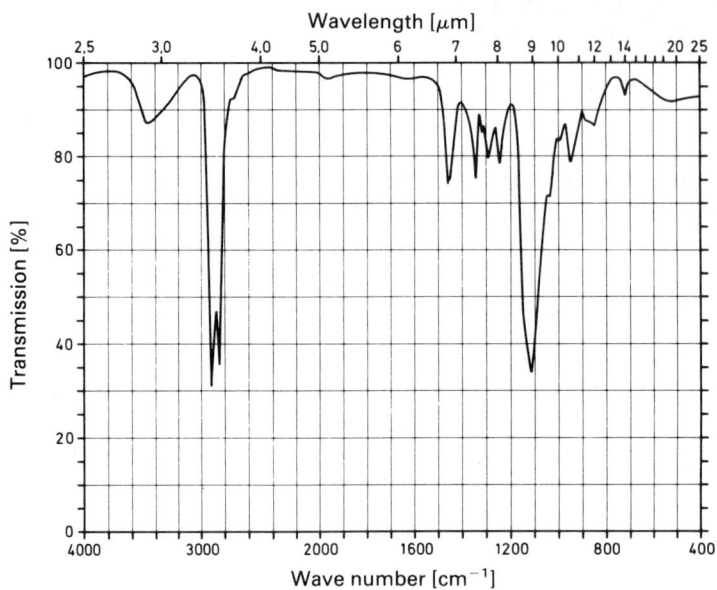

Fig. 12.11 — Scanning spectrum of a linear alcohol ethoxylate on KBr (alkyl=C_{16}, 10×EO).

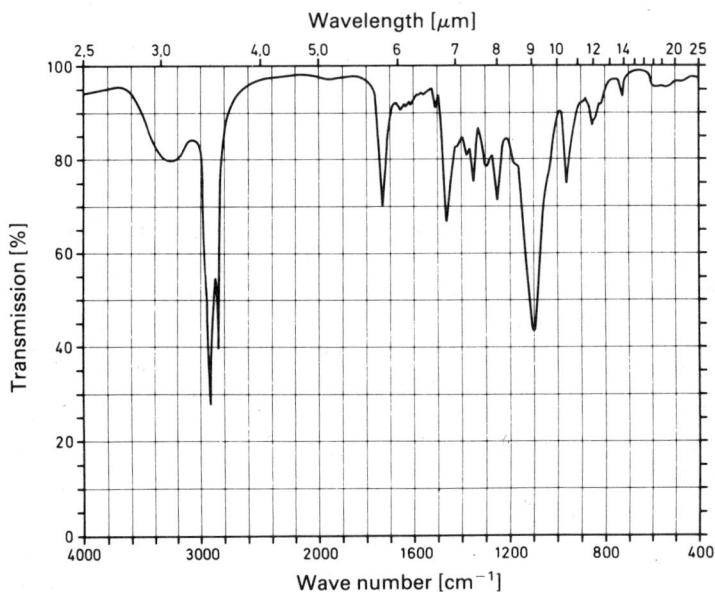

Fig. 12.12 — Biogenic BiAS (pseudo-nonionics) in spring water, after precipitation with
Dragendorff reagent on KBr.

Alkylphenol ethoxylates can, however, be recognised with the aid of the aromatic bands referred to earlier. The concentration is then approximately measurable, such that in the middle and lower Rhine nonionic detergent contents of the order of 0.01–0.02 mg/l have been recorded. The same problems are involved in the determination of nonionics in sludges.

In view of the currently rather unsatisfactory state of detergent analysis in the concentration range of ≤ 0.1 mg/l, as already indicated, other methods have been explored. Occasionally the so-called potassium picrate method (Favretto *et al.*, 1983) is recommended. In this method, analogous to the BiAS method, the ether oxygen linkage of the alkylene oxide groups is rendered cationic with K^+ (instead of Ba^{2+}) and an ionic complex with the picrate ion (in place of the $[BiI_4]^{2+}$ ion) is produced. The spectrophotometric terminal determination is extremely sensitive, and cationic detergents are reputed not to interfere. The basic deficiency lies rather in the accompanying natural interfering materials, which similarly give rise to an 'onium' complex structure.

12.5 ANALYSIS OF CATIONIC DETERGENTS

The 1982/83 output of 20 000 t/annum of distearyldimethyl-ammonium chloride quoted by the manufacturers is also very relevant to surface water analysis from a quantitative aspect. From practical experience, however, the extraordinary affinity of the cationic detergents for solid surfaces — of whatever form — means that, unlike the other 'true' detergents, their concentration in the water phase is likely to be vanishingly small. The method of determination (disulphine blue method; Waters and Kupfer (1976) and Kunkel (1976)) originates once again from the particular requirements. The anionic dye disulphine blue (DSB) reacts with the detergent cation $[R_4N]^{2+}$ giving a blue complex soluble in $CHCl_3$. Anionic detergents interfere and must be removed beforehand by ion-exchange. As the cationic 'detergents' are marked by a much lower activity at the water/air interface relative to anionic detergents, the latter are sometimes added prior to air-stripping, and also serve to inhibit losses due to adsorption at the surface of the collection flasks.

In the field of surface water analysis, virtually no other results are available besides those of Hellmann (1984). The reason for this, in addition to the difficulties of analysis, is probably the fact that they are more extensively eliminated during the sewage treatment process than either the anionics or the nonionics. In the case of direct discharges to surface waters, the cationic detergents will be almost instantly adsorbed onto the suspended solids. Under these conditions one would expect to find concentrations of cationic detergents below 0.01 mg/l. Fewer problems arise in the analysis of domestic and municipal sewage in the sewage works environment.

As with the analysis of anionics in sludges, the classic methods of detergent analysis are not necessarily directly applicable to sludges. Sewage sludge and sludges from natural waters as well as suspended sediments contain large amounts of mostly natural organic compounds, the management and separation of which by ion exchange methods can still not be satisfactorily accomplished. Certainly for the cationic detergents, with their very high adsorptive tendencies, the study of the solids is of much greater importance than that of the water phase. Hence attempts have been made to modify the method so that the use of the customary ion exchanger can be avoided. Actually DSDMAC in aqueous solution can be converted into the cationic $-Ba[BiI_4]_2$ complex using the modified Dragendorff reagent, and this is then extracted into CCl_4. Anionic detergents do not react, but nevertheless migrate to some extent into the organic phase. The extinction of the yellow-orange complex is measured, after transfer to $CHCl_3$ solution at 500 or 490 nm (Hellmann 1984). Nonionic detergents are similarly precipitated at first but are then insoluble in CCl_4. They can thus be removed by filtration and determined as described in Section 12.4.

Derivatives of imidazolines are insoluble in CCl_4 and hence not capable of determination in this way The dodecylbenzyl-dimethylammonium chloride used as a disinfectant, owing to its much lower concentration (by some orders of magnitude), is not normally measurable within the customary framework of detergent analyses.

The very simple method of direct determination of cationic detergents using the Dradendorff complex may in the case of sludges lead to inflated results. It is evident that biogenic BiAS compounds are also partly soluble in CCl_4 (Hellmann 1985). By percolation through Al_2O_3 the cationic detergent fraction may in such situations be cleaned up and the excess Dragendorff reagent simultaneously removed. Following this the DSDMAC may be converted using disulphine blue into the 11 times more sensitive blue complex and determined photometrically at 627 nm.

As in the case of the nonionic detergents the $Ba[BiI_4]_2$ does not interfere with the IR scanning spectrum (Fig. 12.13). In this manner it can also be shown that biogenic

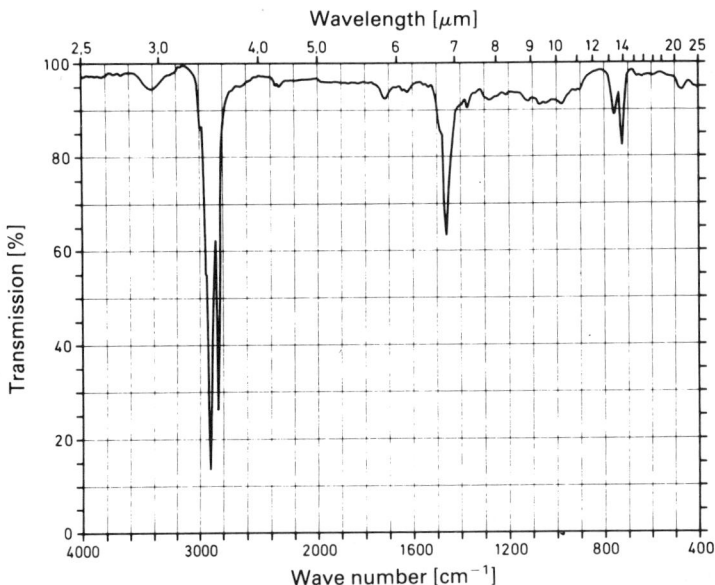

Fig. 12.13 — Scanning spectrum of cationic detergent (DSDMAC) from a sewage sludge sample after precipitation with Dragendorff reagent on KBr.

BiAS are also included in the result (Fig. 12.14). If one continues the analytical

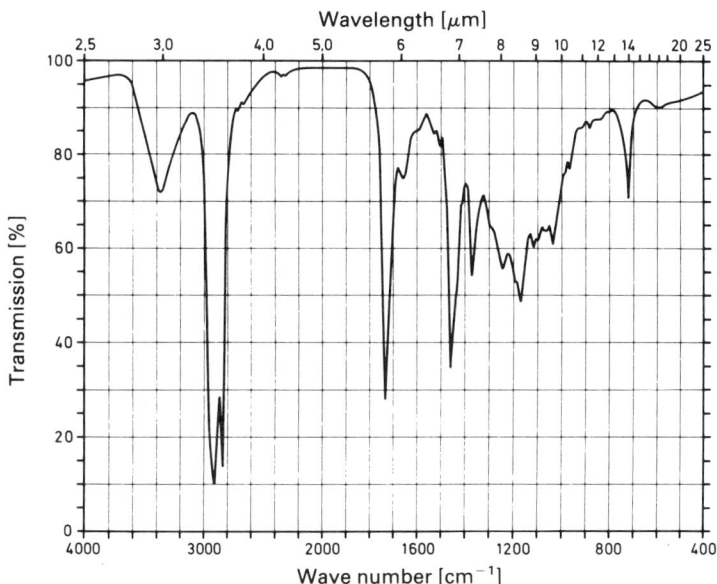

Fig. 12.14 — Scanning spectrum of a mixture of cationic detergent (DSDMAC) and biogenic BiAS from a sample of sewage sludge after precipitation with Dragendorff reagent on KBr.

process using the DSB-complex, then the DSB anion must be separated prior to the IR spectroscopic analysis. This is done by ion-exchange on silica gel in a similar fashion to that for the anionic detergent –methylene blue complex (Hellmann 1983 and p. 170).

In closing a comment on the extraction of detergents from sediments and sludges should not be omitted. All three types can be extracted using methanol from dried, heavily polluted samples. The cationic detergents bound to clay particles may be extracted (Weiss 1982) in the course of the equilibration process using a mixture of $CH_3OH/C_2H_5OH/CHCl_3/6N$ HCl in equal proportions by volume (Hellmann 1984).

12.6 CLOSING REMARKS

From a critical appraisal of the methods of detergent analysis in natural water it may be concluded that there is still considerable opportunity for improvement with respect to all three groups. As the detergent concentrations usually reach only very low values, the separation of the frequently overriding interfering substances is essential. For this the methods previously devised in connection with the manufacture and use of detergent products cannot be employed without more or less extensive modifications, should they be required in the context of surface water analysis. Also the points of view of the 'pure' detergent analyst and the surface water analyst are different. The former concerns himself with what is actually produced, and the latter with what is still detectable in the surface water sytem.

The following tasks remain to be solved (Kunkel 1983): the determination of non-methylene-blue-active anionic detergents and the quantitative estimation of non-bismuth-active detergents (alkanolamides, low-ethoxylated compounds, PO-adducts, etc.).

LITERATURE

12.1

DIN 38409 Teil 23 (Mai 1980). Bestimmung der methylenblauaktiven und der bismutaktiven Substanzen (H 23).

Waters, J., Kupfer, W. (1976). The Determination of Cationic Surfactants in the Presence of Anionic Surfactant in Biodegradation Test Liquors, Anal. Chim. Acta **85**, 241–251.

Wickbold, R. (1971). Anreicherung und Abtrennung der Tenside aus Oberflächengewässern durch Transport in der Grenzfläche Gas/Wasser, Tenside **8**, 61–63.

Die Analytik der Tenside (1976). Monographie der Chemischen Werke Hüls, AG, Marl.

Kunkel, E. (1983). Zum Stand der Tensidanalytik für Wasser, Abwasser und Schlamm, Vom Wasser **60**, 49–58.

Internationale Kommission zum Schutze des Rheins gegen Verunreinigung (ab 1965). Zahlentafeln der physikalisch-chemischen Untersuchungen des Rheins sowie der Mosel/Koblenz.

12.2

Fischer, W. K. (1980). Entwicklung der Tensidkonzentrationen in den deutschen Gewässern 1960–1980, Tenside Detergents **17**, 250–261.

Hellmann, H. (1978). Aniontenside im Rhein 1971–1977. Tensidabbau und Abbaupotenz im Strom, Tenside Detergents **15**, 291–294.

12.3

Hellmann, H. (1983). Kieselgelschichten als Ionenaustauscher bei der Tensidanalytik, Fresenius' Z. Anal. Chem. **315**, 612–617.

Hellmann, H. (1978). Nachweis und Bestimmung von Aniontensiden in Gewässern und Abwässern durch IR-Spektroskopie, Z. Anal. Chem. **293**, 359–363.

Hellmann, H. (1979). Analytische Bestimmung von Aniontensiden in Schweb- und Sinkstoffen sowie Klärschlämmen, Fresenius' Z. Anal. Chem. **295**, 393–397.

Hellmann, H. (1979). Zur Bestimmung von nichtionischen Tensiden in Fluß- und Abwasser durch Röntgenfluoreszenz und IR-Spektroskopie, Fresenius' Z. Anal. Chem. **297**, 102–106.

Hellmann, H. (1985). Pseudo-Niotenside in Gewässern, Vom Wasser **64**, 29–42.

Favretto, L., Stancer, B., Tunis, F. (1983). An Improved Method for the Spectrophotometric Determination of Polyoxyethylene Non-Ionic Surfactants in Waters as Potassium Picrate Active Substances in Presence of Cationic Surfactants, Intern. J. Environ. Anal. Chem. **14**, 201–214.

12.4 and 12.5

Kupfer, W. (1982). Spurenanalytik von kationischen Tensiden unter den speziellen Bedingungen im Wasser und Abwasser, Tenside Detergents **19,** 158–161.

Waters, J., Kupfer, W. (1976). The Determination of Cationic Surfactants in the Presence of Anionic Surfactant in Biodegradation Test Liquors, Anal. Chem. Acta **85,** 241–251.

Kunkel, E. (1976). Bestimmung der kationaktiven Tenside mit Disulfinblau, Hüls-Monographie "Die Analytik der Tenside", S. 102–105.

Hellmann, H. (1984). Einfache spektrophotometrische Bestimmung von k-Tensiden in Gegenwart von a- und n-Tensiden, Fresenius' Z. Anal. Chem. **319,** 272–276.

Hellmann, H. (1986). Zur spektrophotometrischen Bestimmung von k-Tensiden-ein Nachtrag, Fresenius' Z. Anal. Chem., **323,** 29–32.

Weiss, A. (1982). Die Adsorption kationischer Tenside an Mineraloberflächen, Tenside Detergents **19,** 157.

Hellmann, H. (1984). Remobilisierung und Bestimmung von k-Tensiden in Tonmineralien, Fresenius' Z. Anal. Chem. **319,** 267–271.

13

Other substances and groups of substances

13.1 PRELIMINARY REMARKS

Besides the substances and groups of substances referred to in the preceding sections, there are still others which seem worthy of mention in the context of surface water analysis. Especially in recent times, new parameters have constantly been introduced, such as the PCBs and the PAHs, quite apart from the 129 substances listed by the EEC Directive 76/464. Before very long the polychlorinated triphenyls, the dibenzodioxins and the dibenzofurans (also polychlorinated) could be included. It has long been recognised, however, that only special laboratories are fully competent to undertake the tasks of analysis in the ultra-trace analysis region. This specialisation will in the end make it impossible to determine the entire repertoire of classical parameters and modern trace constituents simultaneously in *one* laboratory. If therefore for reasons of environmental politics, numerous official laboratories are obliged to extend their capabilities, then this will entail not only a wasteful division of effort, which should not be underestimated, but also an objective risk concerning the reliability of the analytical results. The results of several ring tests to date among different establishments point in this direction. The following selection of noteworthy parameters is arbitrary but, until section 13.7 is reached, is concerned with familiar parameters of old or more recent origin. They are not dealt with in the fullest sense, as in Chapters 6–12, but rather more briefly with particular aspects in mind. The particular point of emphasis may be concerned with either the analytical techniques, their interpretation or the hydraulic evaluation of the results. The inferences should always be considered in the light of the contents of the earlier chapters.

Where results have been drawn from the specialist literature, these have generally been revised and adapted to the special form of presentation adopted in this monograph.

13.2 READILY CONDENSED SUBSTANCES—PHENOLS

That industrially occurring phenols, and especially the chlorophenols, give rise to taste and odour problems in water and fish flesh, quite apart from their potentially toxic properties, is incontestable. For the surface water analyst there arise, as in the case of the detergents, questions regarding the reliability of the method of determination. If the results of 14-day spot samples during the period 1954 to 1976 (Fig. 13.1) and also those for the Rhine at Koblenz are examined, then the reported values are

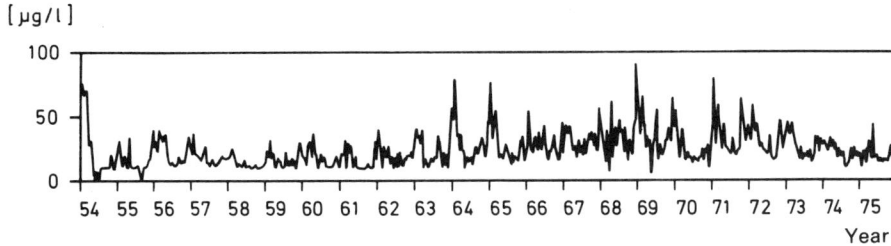

Fig. 13.1 — So-called phenols (readily condensed substances) in the Middle Rhine at Branbach, 14-day spot samples.

mainly below 0.05 mg/l. Particularly low values were reported from 1955 to 1963. They coincided with relatively low water flowrates. High flowrates such as those in 1970 are, however, not characteristically associated with particularly low concentrations of the materials, as would be expected on the basis of the dilution effect. The transport/discharge relationship is on the whole not conclusive. As phenol decomposition in the Rhine must be taken into account, the water temperature should not be omitted from a precise hydraulic evaluation.

In the attempt to determine specific individual phenols of known chemical composition by gas chromatography and to compare the sum total of these results with the result of spectrophotometric measurement (Fig. 13.2), differences between

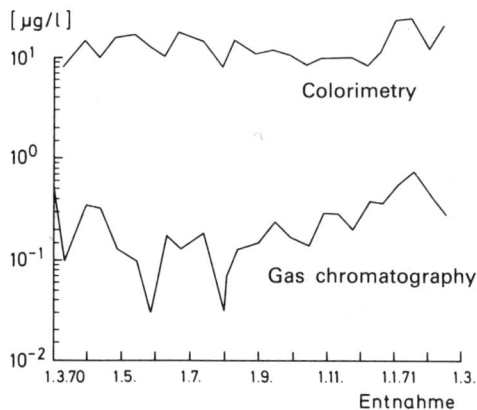

Fig. 13.2 — Phenols in the Main at Schweinfurt, 1970/71. Difference between colorimetric and gas-chromatographic methods of determination (after Kunte, 1971).

the two sets of values of more than an order of magnitude may be apparent. The findings illustrated in respect of the Main at Schweinfurt in 1970/71 can be translated in their entirety to the Middle Rhine (Section 13.5) and possibly also to other watercourses as well. As a rule, nowadays, phenolic compounds of natural origin are by far the dominant fraction. As a result the analytical method does not only lack in precision, but also in its 'screening function'. Of course the monitoring function in respect of certain industrial or trade discharges is not necessarily affected, while an assessment of water quality by means of the phenol-index may in any case still be of local significance.

13.3   DISSOLVED ORGANIC CARBON (DOC)

The dissolved organic carbon content is a very important parameter in connection with the treatment of surface waters for potable purposes. The DOC value has for many years been used by the waterworks situated on the Rhine, the ARW and IAWR. The limiting values stipulated in the IAWR memorandum are 4 mg/l for Group A and 8 mg/l for Group B.

Although at any given point a fall in concentration with increasing discharge may frequently be observed, as shown by Fig. 13.3 for 1973/74, which takes on a

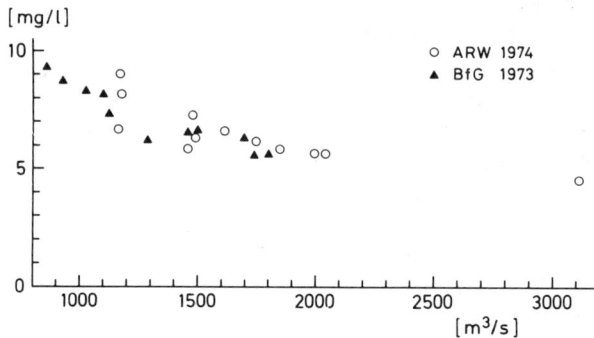

Fig. 13.3 — DOC concentrations in the Rhine at Koblenz as a function of discharge.

hyperbolic form, when the values are converted to DOC transport rates, an increase in transport with an increase in discharge is the normal rule. Opinions in the literature (and in the ARW reports) are agreed that the DOC content of surface waters in such cases can only be partly attributed to effluent discharges. A further portion will be introduced with the runoff from built-up and non-paved areas, either in the dissolved state initially, or by dissolution from the suspended solids—see Figs 4.3 and 4.4. In addition, there will be contributions from the solubilisation of sediments and sludges entrained in the flow, and also the initial contamination in the rainfall itself. An unduly close correlation between the DOC content and the simultaneous discharge value is also inhibited by the temperature-dependent synthesis and decomposition processes of the microbial flora.

Another type of presentation of the results of surface water analyses which should not be omitted is the comparative illustration of the DOC pollution level for several different watercourses on a common graph (Fig. 13.4). The statistics for a

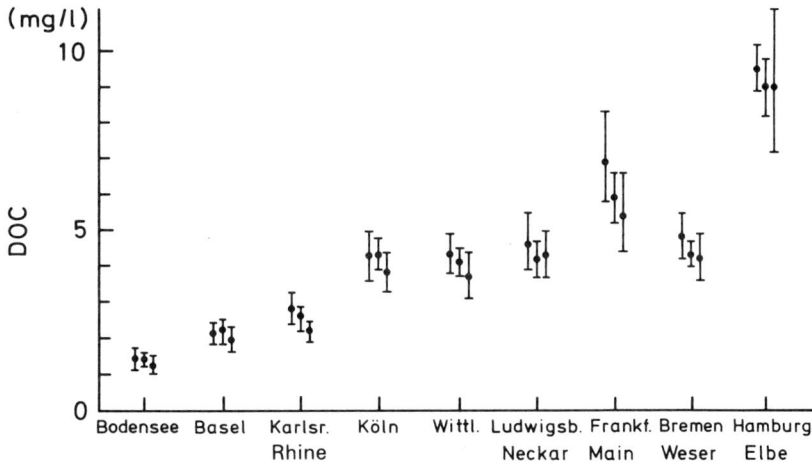

Fig. 13.4 — Range of variation of DOC contents in different rivers during 1976–78, from Sontheimer (1981), one sampling point in each case; years 1976/77/78 (l. to r.).

given sampling point shown on the graph comprise the upper and lower limits for standard deviation about the arithmetic mean. This highly interesting form of presentation, even though no account is taken of the discharge, is the more informative when sufficient measurements, covering a wide range of discharge values, are available. With this proviso it enables the different degrees of pollution of the rivers in question to be assessed, as well as the increase in DOC pollution level for a particular river in the downstream direction. Should extreme values be observed in the wake of very high flowrates, their inclusion in the picture is not always meaningful. According to Fig. 13.4 the DOC content of Lake Constance, for example, is between 1 and 2 mg/l while in the Rhine at Karlsruhe, values between 2 and 3 mg/l were recorded (1976/78). Since 1973/74 the DOC pollution level of the Middle and Lower Rhine has diminished, as can readily be seen from a comparison of Figs 13.3 and 13.4.

The values for DOC content determined from time to time as part of measurement programmes undertaken by the ARW, the IARW and latterly the IKSR are very difficult to evaluate on hydraulic grounds and to that extent are somewhat irrelevant. The problem manifests itself chiefly in the results for combined samples from automatic sample collectors, where very often the nature and composition and also the amount of suspended solids in the sample do not correspond with the situation in the main body of the river.

13.4 LIGNINSULPHONIC ACIDS AND HUMIC ACIDS

Of the total flux of dissolved organic matter, a considerable percentage may consist of ligninsulphonic acids. There is evidence that they originate from industrial sources. The large fluctuations in the quantities determined by analysis in the Rhine (Fig. 13.5) are an indication of the relatively irregular nature of the inputs.

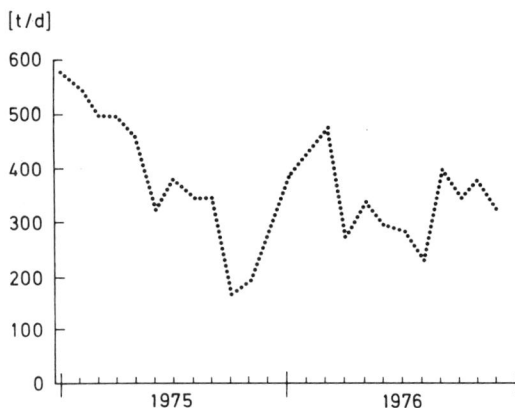

Fig. 13.5 — Lignin sulphonic acid transport in the Middle Rhine at Koblenz, from Sontheimer (1977).

According to Sontheimer the mass flux is largely independent of discharge. The extensive effluent treatment facilities introduced by the relevant industrial dischargers of recent times have enabled the transport rate in the Rhine to be substantially reduced, as shown by the data summaries for 1976, 1977 and 1978 in Fig. 13.6.

The chemical grouping of ligninsulphonic acids is of particular interest in the analysis of surface waters because they originate from a very limited number of industrial premises and their determination in surface waters is not complicated by the presence of materials of biogenic origin, as in the case of the MBAS and BiAS (Chapter 12). To this may be added the fact that they are only slowly degraded, compared with the customary flow times for inland waterways on their way to the sea.

The humic acids which originate from natural sources are also very poorly degradable. It may even be a fact that further substances of this kind are formed during the time of transport from inland districts to the sea. As suspended solids and sediments are rich in humic acids, any change in suspended solids transport will lead to a change in the transport of humic acids. As long as the entrainment of sludges and sediments transmits further quantities of dissolved humic acids into the main stream, their concentration should increase along with an increase in discharge. The data concerning mass transport rates as a function of discharge summarised by Sontheimer (1977) provide confirmation of this hypothesis. The scatter of individual

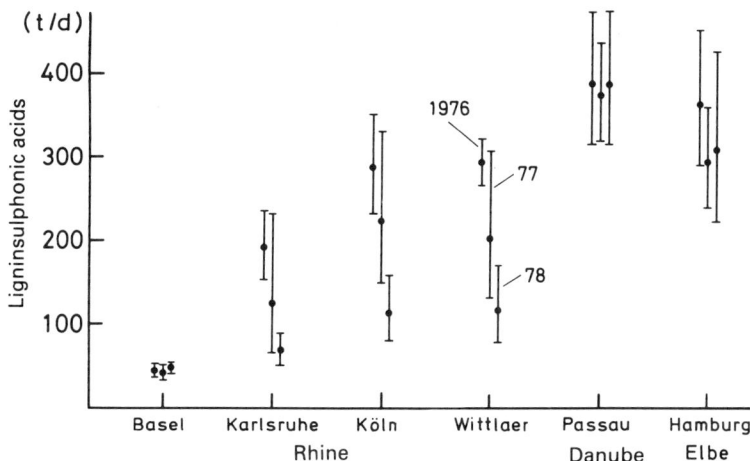

Fig. 13.6 — Ligninsulphonic acids in different rivers. Range of variation in mass transport rates, 1976–78 (Sontheimer, 1981).

values for equal values of discharge (Fig. 13.7) should be explicable in terms of the circumstances explained in detail in Chapter 6, which occur during rising and falling flood flows, and give rise to a pear-shaped profile for the concentration vs discharge relationship (Fig. 6.14). Thus the flux increases disproportionately as the discharge approaches its peak, while as the flow subsides the values tend to fall below the equilibrium curve passing through the origin of coordinates. The general order of magnitude for the Middle and Lower Rhine is around 1.7 mg/l.

In certain areas of North Germany greater importance may be attached to the presence of humic acids than in the Rhine basin, as for example on the Lower Ems and the Weser, where surface waters contribute directly to the production of potable supplies. Their role in the production of haloforms in the course of the preparation of drinking water may be noted in passing. The binding of heavy metals by humic acids is also of scientific interest.

13.5   ORGANOCHLORINE COMPOUNDS OF INTERMEDIATE BOILING POINT

In chapter 10 some of the problems connected with high-boiling chlorinated hydrocarbons and in Chapter 11 those of the low-boiling chlorinated hydrocarbons were considered. Besides these groups of compounds, certain laboratories have been in the habit for the past 10 years or more of looking for certain chlorinated aromatics, which were later incorporated with the 129 substances listed in the 76/464 EEC Directive. In particular these consisted of chlorinated benzenes and substituted chlorobenzenes such as chloronitrobenzene, chloroanilines and chlorophenols. If the results of previously published surface water studies such as those of Borneff *et al.* (1978) are considered, it will be apparent that some chlorobenzenes invariably

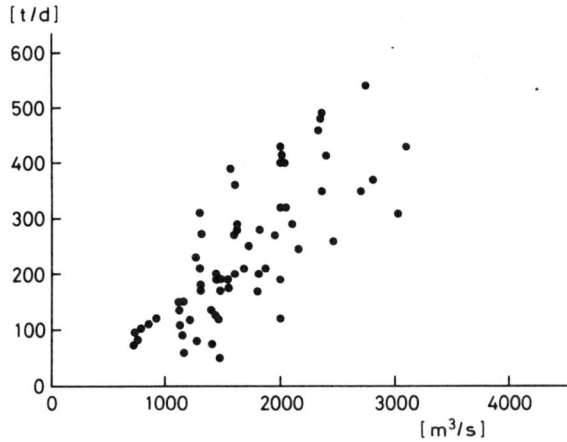

Fig. 13.7 — Mass transport of humic acids in the Rhine as a function of discharge data from ARW (1974), for various sampling points.

occur where sewage effluents are discharged into surface waters. Particularly prominent are the 1,2-dichloro- and 1,4-dichlorobenzenes. The occurrence of the latter is doubtless occasioned by the use of toilet disinfectants.

In contrast to this, the three chloronitrobenzene isomers are found only in certain stretches, so that it can be inferred that they originate from a small number of point-source industrial discharges. The concentration range for both the dichlorobenzenes and the chloronitrobenzenes in the Rhine basin lies principally in the region 0.1–5 μg/l, sometimes more and sometimes less. Among the chlorophenols, pentachloro-phenol is of particular note, its widespread occurrence probably related to its extensive use as a constituent of preservative agents for wood.

A reliable indicator for the origin of certain substances from a localised point source in the broadest sense, including the wash-down of contaminated surfaces, is the large fluctuation in the individual concentration values obtained for equal values of the discharge. These fluctuations may occur within a period of a day or several days depending on the periodicity of the effluent flow and the size of the receiving stream (Fig. 13.8). Where it is suspected that several compounds originate from a single point source, one may obtain firmer indications from a study of correlations between the results. According to Table 13.1 roughly constant relationships were observed between the 2- and 3-chloronitrobenzene isomers and between the 2- and 4-chloronitrobenzene isomers. One may also derive indications in this way of the internal consistency of the analytical results, as already discussed in Section 10.3.

These values for choronitrobenzene obtained by a cooperating laboratory showed a definite concentration peak on 20 September. The 'base load' at this point was very much smaller. The values obtained later in our own laboratory for chloronitrobenzene compounds in the Middle Rhine during 1985 (Table 13.2) were roughly equal to the base load values of Table 13.1.

The concentration of 1,3-chloronitrobenzene was once again the lowest of the

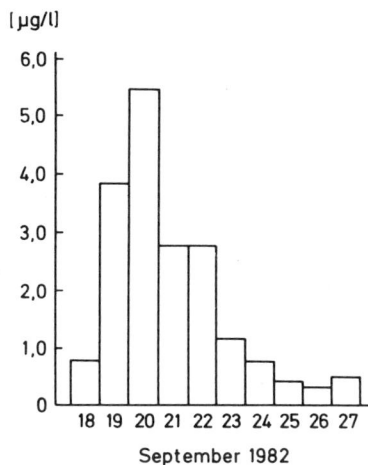

Fig. 13.8 — Chloronitrobenzenes (2-, 3-, and 4-chloro- isomers) in the Rhine at Koblenz; 24 h combined samples.

Table 13.1 — Chloronitrobenzenes (CNB) in the Rhine at Koblenz, for 24 h combined samples

Sampling date 1982	2-CNB (μg/l)	3-CNB (μg/l)	4-CNB (μg/l)	$\dfrac{\text{2-CNB}}{\text{3-CNB}}$ (ratio)	$\dfrac{\text{2-CNB}}{\text{4-CNB}}$ (ratio)
18 Sept.	0.40	0.06	0.33	6.7	1.2
19 Sept.	2.1	0.36	1.4	5.8	1.5
20 Sept.	2.9	0.57	2.0	5.1	1.4
21 Sept.	1.5	0.27	1.0	5.6	1.5
22 Sept.	1.4	0.30	1.1	4.7	1.3
23 Sept.	0.53	0.11	0.51	4.8	1.0
24 Sept.	0.32	0.07	0.37	4.6	0.9
25 Sept.	0.17	0.05	0.21	3.4	0.8
26 Sept.	0.18	<0.05	0.12	<3.6	1.5
27 Sept.	0.26	<0.05	0.24	<5.2	1.1

Table 13.2 — Dichlorobenzenes (DCB) and chloronitrobenzenes (CNB) in the
Rhine at Koblenz, spot samples, in μg/l

| Sampling date | 1,2-DCB | 1,3-DCB | 1,4-DCB | 2-CNB | 3-CNB | 4-CNB | 2-CNB/3-CNB |
1985	(μg/l)	(μg/l)	(μg/l)	(μg/l)	(μg/l)	(μg/l)	(ratio)
25 Feb.	0.65	0.23	0.42	0.20	0.07	0.19	2.8
26 Feb.	1.2	0.66	0.60	0.21	0.08	0.42	2.6
27 Feb.	1.1	0.56	0.64	0.10	0.05	0.28	2.0
28 Feb.	1.5	0.69	1.0	0.14	0.08	0.38	1.8
1 Mar.	2.6	0.54	0.92	0.11	0.06	0.24	1.8
4 Mar.	2.8	0.60	0.82	0.17	0.13	0.30	1.3
5 Mar.	1.4	0.63	0.65	0.08	0.03	0.12	2.7

three isomers. The ratio of concentrations for 1,2-chloronitrobenzene and 1,3-chloronitrobenzene deviates noticeably from the relationship obtaining in Table 13.1. However, as far as can be inferred from the values in the Table, the analyses appeared to be self-consistent. If one were to assume (which is by no means certain) that there had been no change in the composition of the effluents, then the problem of instrument calibration ought to be looked into, in order to eliminate the discrepancy.

The chlorinated aromatics in the medium-boiling range must also be considered from the standpoint of their potential accumulation in sediments and organisms. For the analyst this means that their distribution between the water and suspended solids phases must be investigated in each case. In the second place, the specific contamination load of sludges and sediments must also be determined.

It is possible to state here and now that the potential for enrichment of these materials by solids in the surface water system is several orders of magnitude lower than for other compounds like HCB or DDT.

13.6 DISSOLVED ORGANOCHLORINE COMPOUNDS (DOCl)
Organic chlorine compounds are nowadays detectable in very many surface water systems. Based on their potential toxic hazards it is possible to distinguish several groups of compounds. It should be noted, however, that their aquatic toxicity is not necessarily synonymous with their risk to health.

Chlorinated compounds with a strong tendency to accumulate in sediments and living organisms, like HCB, DDT and PCBs, occur in concentrations less than 0.1 μg/l, and frequently lower than 0.01 or 0.001 μg/l. If one adds up the concentrations of all such substances, then it is rarely possible, even in heavily polluted waters, for the total value to exceed 1 μg/l.

The low-boiling chlorinated hydrocarbons from dichloromethane to perchloroethylene are encountered in higher concentrations. Their sum total, however, would reach or possibly exceed 10 μg/l only on exceptional occasions and in very few river reaches. The chlorobenzenes and substituted chlorobenzenes already referred to in Section 13.5 may *in toto* and in certain reaches amount to 10 or even 20 μg/l and over. Added to these are the chloroligninsulphonic acids, for which in the early 1970s values of several mg/l could be obtained. If one looks mainly at the relative importance of different groups, then the contribution of the ligninsulphonic acids far outweighs that of any other class of compounds.

The analyst on the contrary has no choice but deliberately to analyse for certain compounds. Over and above this, it can be of considerable benefit, both for an assessment of water quality by itself and also as a measure of its usefulness for potable supplies, to determine the total of the organic chlorine compounds as DOCl (dissolved) or TOCl (dissolved and undissolved). Owing to the low accumulation tendency for the more prominent organohalogen compounds in suspended solids, the value of DOCl is normally very close to that of TOCl. In Fig. 13.9 the

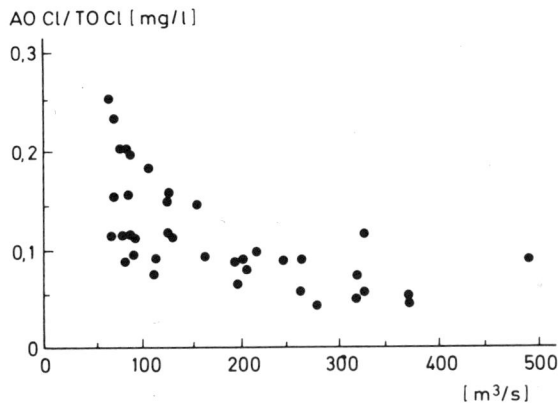

Fig. 13.9 — Combined organic chlorine in the Main at Frankfurt (km 30). Values obtained between 1977 and 1979, without distinguishing between AOCl and TOCl methods (Sontheimer, ARW 1979).

measurements obtained following adsorption onto activated carbon (AOCl) and the values given by a direct determination on the sample (TOCl) are plotted together (after Sontheimer, and ARW 1979). The anthropogenic/industrial origin of these substances manifests itself once again in the hyperbolic decrease of the concentrations with increasing flowrate, while the mass transport rate stays appreciably constant.

The range of scatter of the individual values as well as the specific contamination level of various rivers and stretches of waterway are again well demonstrated in the manner shown in Fig. 13.10.

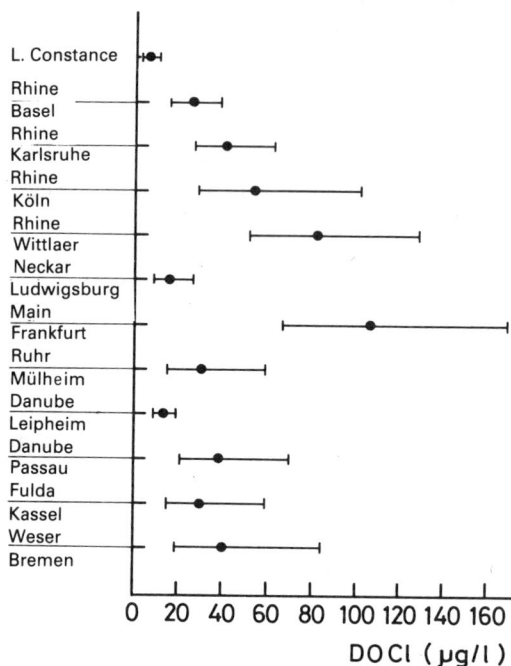

Fig. 13.10 — Range of variation of DOCl contents from 1976 to 1978 for different rivers (Sontheimer, 1981).

Latterly the sum of the organic halogen compounds (AOX or EOX, where X = F, Cl, Br, I) has been commonly reported instead of the customary DOCl (as AOCl or EOCl), not merely for analytical reasons. However, in inland waters the chlorine compounds represent by far the largest proportion of the total halogen derivatives, in our experience.

13.7 DIBENZODIOXINS AND DIBENZOFURANS

Since analyses of refuse deposits, soil samples and exhaust gases, as well as certain technical products like pentachlorophenol and used oils, have revealed the presence of dioxins and furans—strictly speaking, polychlorinated dibenzodioxins (PCDD) and polychlorinated dibenzofurans (PCDF)—the time will probably come when surface water sludges and sediments, and perhaps even the water phase itself, will be involved in this type of determination.

There is still no evidence of any real problem. From the hydraulic aspect, any evaluation of results would follow similar lines to that of the PCB data. If one assumes that not all of the 210 possible members of the PCDD and PCDF categories are environmentally or toxicologically relevant, then the question of the 'correct' choice of a particular target compound arises. Without doubt priority should be given to members of the group of 2,3,7,8-tetrachloro derivatives (Ballschmiter 1985), the well-known Seveso poison. To what extent other individual compounds of similar toxicity merit inclusion on the so-called Black List, remains to be seen.

The case of the PCDD and PCDF compounds certainly casts a new light on the 129-member list of substances drawn up by the EEC. A corresponding adherence to the saying of Goethe (*Faust,* Part 1)—'Whoever brings a lot must bring many things' ('*Wer vieles bringt, wird manchem etwas bringen*')—is no longer feasible in the case of the lone analyst, and specialisation becomes the order of the day.

LITERATURE

13.2
DIN 38 409 Teil 16: Bestimmung des Phenolindex (H 16). Entwurf 1982.
Kunte, H., Slemrova, J. (1975). Gaschromatographische und massenspektrometrische Identifizierung phenolischer Substanzen aus Oberflächengewässern, Z. Wasser-Abwasser-Forsch. **8**, 176–182.
Koppe, P., Traud, J. (1973). Untersuchungen über phenolartige Stoffe in Wässern, Gewässerschutz, Wasser, Abwasser **10**, 345–366.
Scholz, L. (1963). Phenolanalyse in Trinkwasser und Vorfluter, Vom Wasser **30**, 143–152.
Naumann, E., Scholz, L. (1964). Beitrag zur Entstehung kupplungsfähiger Stoffe durch Zersetzung biologischer Materie, Gas- und Wasserfach **105**, 707–709.
Selenka, F. (1972). Schadstoffbelastung unserer Oberflächengewässer, Veröffentlichungen des Instituts für Küsten- und Binnenfischerei, Hamburg (S. 17, Darstellung nach H. Kunte 1971).
Dietz, F., Traud, J. (1978). Zur Spurenanalyse von Phenolen, insbesondere Chlorphenolen in Wässern mittels Gaschromatographie – Methoden und Ergebnisse, Vom Wasser **51**, 235–257.

13.3
Sontheimer, H., Gimbel, R. (1977). Untersuchungen über die Veränderung der Fracht an organischen Wasserinhaltsstoffen mit der Wasserführung am Beispiel des Rheins, GWF-Wasser/Abwasser **118**, 165.
Rheinberichte der Internationalen Arbeitsgemeinschaft der Wasserwerke im Rheineinzugsgebiet; z. B. 1979, S. 69, Abb. 8 sowie 1977, S. 48, Abb. 2.
Arbeitsgemeinschaft Rheinwasserwerke e.V.; z. B. Jahresberichte 1974 u. 1979, Tabellen im Anhang.
Berichte der "Rijncommissie Waterleidingbedrijen" der Niederlande.
Sontheimer, H., Kühn, W. (1981). Organische Schadstoffe in den Fließgewässern der Bundesrepublik Deutschland, DVGW-Schriftenreihe – Wasser Nr. 26, Frankfurt am Main.
Sontheimer, H., Fuchs, F., Kühn, W. (1978). Beurteilung der organischen Belastung von Fließgewässern, Vom Wasser **51**, 1–10.

13.4
Sontheimer, H. (1977). Neuester Stand der Rheinverschmutzung, IAWR 6, S. 54.
ARW (1974). 31. Ericht, S. 79–91.
Umweltprobleme des Rheins. 3. Sondergutachten 1976, W. Kohlhammer Verlag GmbH, Stuttgart u. Mainz, S. 58.

Sontheimer, H., Wagner, J. (1977). Zur Bestimmung von Huminsäuren und Ligninsulfonsäuren aus den UV-Spektren, Z. Wasser-Abwasser-Forsch. **10**, 77.
Sontheimer, H., Kühn, W. (1981). See under 13.3.

13.5

Borneff, J., Hartmetz, G., Fischer, A. (1978). Die Belastung von Oberflächenwasser mit Chlorbenzolen, Wasser Forschungsbericht 14/74 für das Bundesministerium des Innern.
Fischer, A., Slemrova, J. (1978). Die Belastung des Rheins mit Chlorbenzolen, Vom Wasser **51**, 33–46.
Mitteilungen der Landesanstalt für Wasser und Abfall Nordrhein-Westfalen. Jahresberichte und Sonderuntersuchungen, z. B. Heft Oktober 1977 bis März 1978.
Wegmann, R. C. C., van den Broek, H. H. (1983). Chlorophenols in River Sediment in the Netherlands, Water Res. **17**, 227–230.
Augustin, H., Bauer, U. u. Arbeitsgruppe (1982). Mikrobiozide Wirkstoffe als belastende Verbindungen im Wasser, Vom Wasser **58**, 297–340.

13.6

Sontheimer, H., Schnitzler, M. (1982). EOX oder AOX?–Zur Anwendung von Anreicherungsverfahren bei der analytischen Bestimmung von chemischen Gruppenparametern, Vom Wasser **59**, 169–179.
DIN 38409 Teil 8: Bestimmung der extrahierbaren organisch gebundenen Halogene (EOX) (H8).
Sontheimer, H., Kühn, W. (1981). See under 13.3.
DIN 38409 Teil 14: Bestimmung der adsorbierbaren organisch gebundenen Halogene (AOX) (H14).
ARW Jahresbericht 1979, S. 80 u. a.

13.7

Ballschmiter, K. (1985). Chemie und Analytik der Polychlordibenzodioxine (Dioxine) und Polychlordibenzofurane (Furane), VCI-Schriftenreihe Chemie und Fortschritt **1**, 8–12.
GDCh Presse-Information Erklärung zu Pentachlorphenol (PCP), Frankfurt am Main 28.3.84; dort Angaben zum Gehalt des PCP an Dioxinen.
Zoller, W., Schäfer, W., Class, T., Ballschmiter, K. (1985). Quantitation of Polychlorodibenzodioxin and Polychlorobiphenyl Standards by Gas-Chromatography-Flame Ionisation Detection, Fresenius' Z. Anal. Chem. **321**, 247–251.
Smith, R. M., O'Keefe, P. W. et al. (1983). 2,3,7,8-Tetrachlorodibenzo-p-dioxin in sediment samples from Love Canal storm sewers and creeks, Environ. Sci. Technol. **17**, 6–10.
Wong, A. (1982). Determination of TCDD (2,3,7,8-Tetrachlorodibenzo-p-dioxin) in industrial and municipal waste waters, U.S. Environ. Prot. Agency, Off. Res. Dev., EPA-600/4-82-028, 51 S.
Human and Environmental Risks of Chlorinated Dioxin and Related Compounds (1983). Proceedings of an international symposium on Chlorinated Dioxins and Related Compounds held in Arlington, VA, October 25–29, 1981, Plenum Press, New York, pp. 833.

"Dioxin in der Umwelt" (1985). VCI-Schriftenreihe Chemie und Fortschritt 1.

Umweltbundesamt (Hrsg.) (1985). Der Bericht 5/85 – Sachstand Dioxine, Erich Schmidt Verlag, Bielefeld.

14

Cross-connections

14.1 INTRODUCTION

Because for the purpose of modern trace analysis a subdivision of human resources appears inevitable, just as for the classical analysis of surface waters, a continuous contact between different analytical working groups is all the more essential. Where other subject areas of the science of natural waters are involved as well, the network becomes even larger.

As already explained in the earlier chapters, certain relationships normally exist between the analytical and 'non-analytical' parameters. A knowledge of such relationships or connections can be of benefit on the one hand for checking the validity of the original results, and on the other for a further interpretation of their significance. Besides the often very close connections between the level of a particular quantity in terms of concentration, and the discharge, the water temperature and (depending on the parameter) the suspended solids transport (Part I), and hence between the chemical and the hydrological parameters, numerous other relationships between the purely chemical constituents of natural waters are also recognisable.

One can now obviously distinguish between several types of relationship. A relatively close relationship is to be expected if two different analytical procedures are concerned with roughly the same target quantity, e.g. in the case of the loss on ignition and the organic carbon content of a sediment sample. Not infrequently the moisture content and the loss on ignition of sludge samples can also be correlated. Where two or more constituents originate from the same point source, one may likewise discover useful relationships between their respective concentrations in the water. Moreover, certain substances are often useful as 'tracers', as indicated for the phosphate anion in Tables 8.2 and 9.3. Finally, one is always finding new correlations between the trace substances of organic or inorganic origin and other 'associated' elements.

It is true that in many cases the relationship is not very close and in particular not generally applicable; frequently it cannot be translated from one sampling point to another, especially in respect of sediments. There is certainly no justification for placing such confidence in supposedly sound correlation coefficients that the analysis can be performed at a desk! Experience and expert knowledge are just as essential in the use of cross-correlations as they are in the analytical task itself.

14.2 COGNATE PARAMETERS

Dissolved organic carbon (DOC) and Total Organic Carbon (TOC) are undoubtedly cognate parameters which are determined using the same method. In surface waters of relatively low suspended solids concentration the undissolved carbon is of very little practical importance and the values for DOC and TOC are practically identical. This obtains in the Saar (Fig. 14.1, concentration range 2–7 mg DOC/l) and also for the Rhine and other major rivers under conditions of dry weather flows. For higher suspended solids concentrations such as obtain in heavily polluted receiving streams (Fig. 14.1 — values over 7 mg DOC/l) and in general in the case of

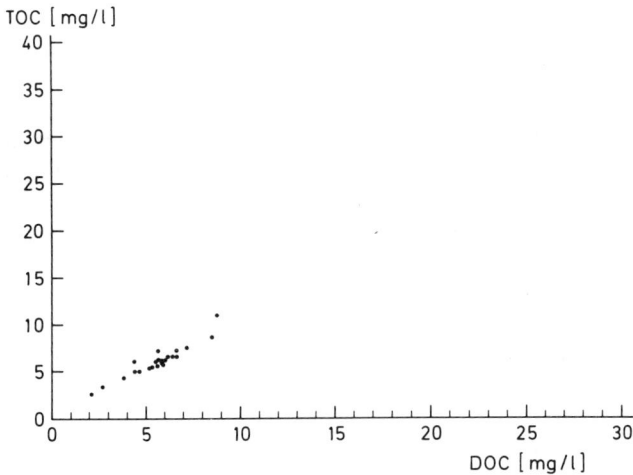

Fig. 14.1 — Relationship between DOC and TOC for water samples from the Saar and tributaries, December 1984.

approaching flood flows, the amount of undissolved carbon reaches noticeable proportions. In exceptional cases the carbon contained in compounds of the CN^- and CNS^- ions may also be included in the result. Although no stoichiometric ratio between the DOC and the TOC and the total quantity of organic matter has been demonstrated, an empirical factor of 2.5 has generally been found to be applicable in moderately polluted rivers.

While the organic carbon content is something which can be clearly defined, the same is not true of the chemical oxygen demand (COD) in which not only organic compounds of varying degrees of oxidation but also certain inorganic compounds are included. According to Fig. 14.2 the heavily polluted tributaries of the Saar fall out of

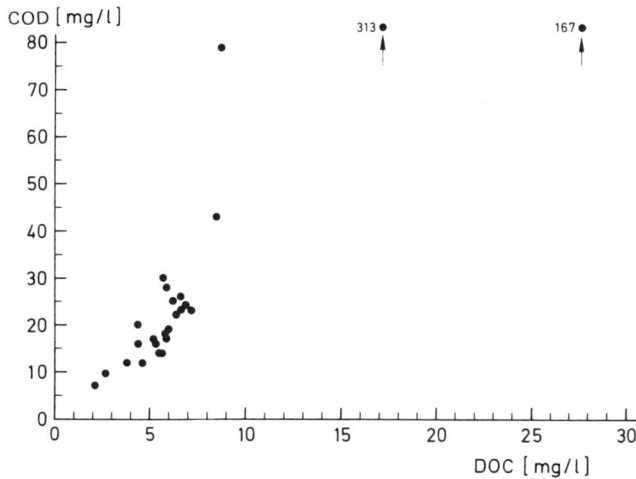

Fig. 14.2 — Relationship between DOC and COD for water samples from the Saar and tributaries, December 1984.

line when one attempts to correlate the COD and TOC. For effluents discharged from sewage treatment plants a factor of 2.95 is normal. According to Sontheimer (1970), during the period from March to December 1969 the corresponding ratios for the Alpenrhein were 1.6, Hochrhein 1.8, Upper Rhine 2.3 and Middle and Lower Rhine 2.4–2.7. Actually a conversion of organic C into oxygen equivalents gives a figure of 2.67. In the Saar, for which according to Fig. 14.2, DOC and TOC appeared to be roughly equal, even for relatively low DOC concentration of 6–7 mg/l, markedly increased COD values were observed, such that the ratio COD/DOC rose to over 3.0. In three cases it was even greater than 10.0, and especially so because significant amounts of ammonia were included in the COD value. Conversely, relationships deviating noticeably from the norm may be indicative of the presence of special substances in the water, and should thus be the occasion for a deliberate reconsideration.

The loss on ignition of a sediment sample at 600°C is likewise no uniquely defined quantity. This is not merely because of the presence of organic matter in various stages of oxidation, but also because of the 'error' introduced by water of crystallisation of certain compounds, which is not vaporised at 105°C, and also in some cases due to the appreciable decomposition of Ca and Mg carbonates at 600°C. The findings plotted in Fig. 14.3 thus deviate by a far from negligible amount from the expected straight-line relationship. For samples with a loss on ignition of up to 10% of the dry solids, the average value for the ratio of loss on ignition to organic carbon is approximately 3.0. It is true that the relationship may be much closer for a narrowly defined sampling area, e.g. for a dock basin or an impounded reach, so that the loss on ignition is always a useful quanity. Disproportionate amounts of organic C suggest the presence of elementary carbon in the form of coal or soot particles, such as are frequently observed in the catchment area of the Saar.

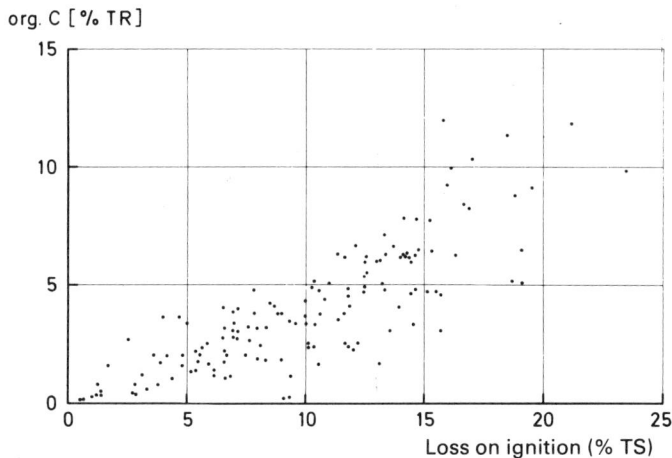

org. C [% TR]

Loss on ignition (% TS)

Fig. 14.3 — Relationship between organic C and loss in ignition for sediments from the Rhine basin. Data for 1964/65.

The analyst will be aware of even more cross-connections. Between the permanganate oxygen demand and the organic carbon ($KMnO_4$/org C) there is often a ratio of 1.5–2.0.

The ratio of COD/BOD_5 in the intake of a sewage works is usually in the range 1.5–2.0 (O_2-equivalents); in the Middle Rhine in 1979 the ratio was between 2.3 and 3.6 and in the Lower Rhine at Bimmen/Lobith between 3.0 and 4.9 (data from the International Commission). Sontheimer also established certain relationships between the UV-extinction coefficients at 240 nm and certain specific water constituents.

14.3 PARAMETERS OF IDENTICAL ORIGIN

Time and again analytical findings lead to the conclusion that two or more substances originate from the same source. The isomeric chloronitrobenzenes (Tables 13.1 and 13.2) occur in certain specific locations in a constant ratio, although at various concentration levels.

Another case in point involves phosphate and anionic detergents, both of which are regular constituents of domestic wastewaters and hence of muncipal sewage. As the anionic detergents, in contrast to the phosphates, are biologically degraded, useful relationships are found in practice only in untreated sewage and in those smaller rivers which are relatively heavily polluted with incompletely treated sewage. This situation was encountered in the Saar and its tributaries, according to Fig. 14.4. However, in the conurbation area of the Lower Main a concomitant increase in the MBAS and total PO_4^{3+} contents was also discernible in the direction of flow (Fig. 14.5). That the MBAS value is practically synonymous with the content of anionic detergents is apparent from the magnitude of the results — see the

MBAS [mg/l]

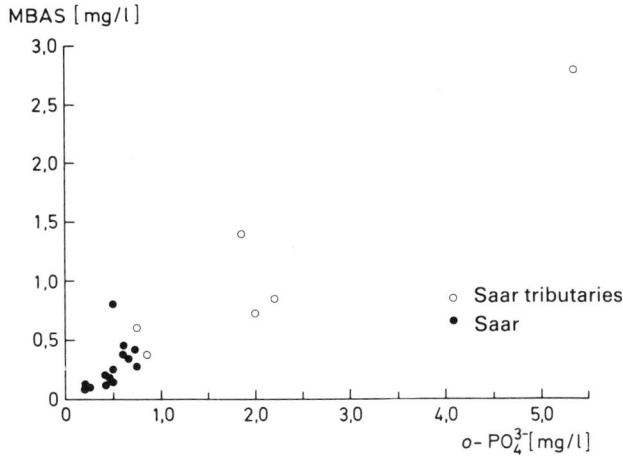

Fig. 14.4 — Relationship between o-PO_4^{3-} and MBAS for the Saar and tributary streams, December 1984.

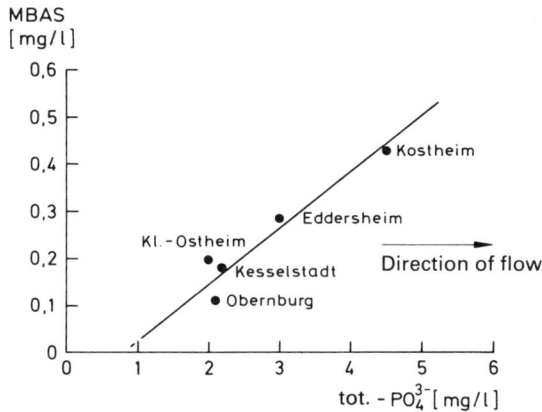

Fig. 14.5 — Relationship between tot. PO_4^{3-} and MBAS for the Lower Main, 1978.

information presented in Section 12. The term 'identical origin' here means simply that the introduction of these materials occurs within a spatially defined catchment area. The not inconsiderable previous contamination of the Main with total PO_4^{3+} at Obernburg can also be discerned in Fig. 14.5.

Sediment profiles are occasionally introduced into this topic after obtaining the analytical results. Thus such a profile was determined for the Saar impoundment at Mettlach, and assessed with respect to the levels of lead and zinc (Fig. 14.6). To an

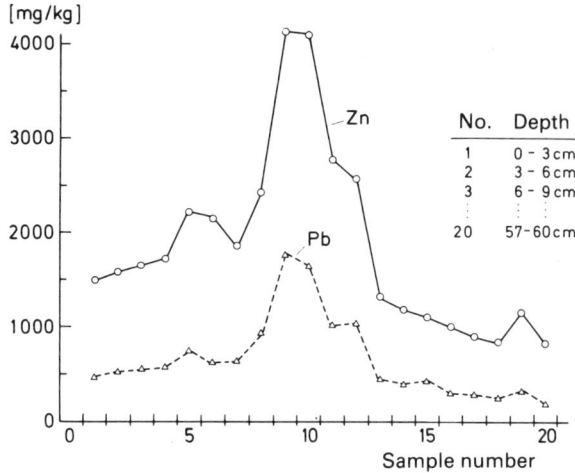

Fig. 14.6 — Common pattern of Pb and Zn concentration profiles as a function of depth for sediment cores obtained form the Metlach impoundment on the Saar, 1977.

experienced analyst the heavy metal concentrations are exceptionally high. The absolutely identical form of the concentration profile for the two elements was compelling evidence that the contamination of such an order with both metals could be assigned to a single discharger. The variable levels of contamination associated with the different sampling positions are not to be regarded as indicative of an irregular release of effluent but rather as a consequence of the sedimentation conditions in this regulated reach, and the varying admixture of solids of a lower level of contamination.

In the analysis of sediment profiles from the Lower Main in 1978/80, particular segments were characterised by greatly increased levels of barium, chromium and zinc, their very close correlation being indicative of a single origin. From time to time one also observes unusually high levels of silver, and sometimes other precious metals, which must have originated from a single, localised source.

The connection between the analytical results and identifiable sources of emission is still most readily demonstrated in smaller rivers of limited catchment area, as long as particularly rare constituents are not concerned. For the most topical heavy metals, for the classical determinands like phosphate and detergents, and for hydrocarbons, polycyclic aromatics and chlorinated organic insecticides such as lindane and DDT, actual pinpointing of the source is not usually possible.

14.4   TRACER PARAMETERS

The phosphate content (total PO_4^{3-}) of spring waters and pristine streams is ordinarily around 0.1 mg/l or below. If one then determines the phosphate content of such a stream below the first human settlement, then in general there is a distinct increase. The phosphate load increases fairly regularly with the 'sewered' population, while the observed phosphate concentration is of course dependent on the actual water flowrate. The phosphate anion becomes for our purposes a tracer for the presence of municipal sewage discharges. Along with the phosphate content there is also an increase in the concentration of other parameters in the aqueous phase, such as the DOC, nitrogen in various combined forms, the BOD and the salt concentration.

With regard to sludges and sediments, similar tracer parameters can be adduced for the potential 'contamination'. The thinking runs as follows: organic and inorganic trace contaminants are prefentially absorbed onto the fine particulate fraction of the sedimentary material, i.e. onto the clay and silt fraction, and onto the dead and animate organic matter. For this reason analytical data in the technical literature are increasingly cited for definitive size fractions, e.g. <0.063, <0.02 or <0.006 mm, or perhaps related to the loss on ignition. A relatively large loss on ignition of the sample is invariably associated with a high moisture content. In Fig. 14.7 the water

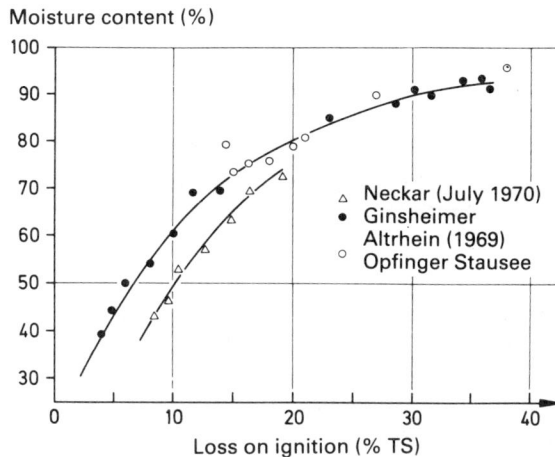

Fig. 14.7 — Relationship between loss on ignition and moisture content in sludge samples.

content as a percentage of the wet weight (% wet wt) is plotted as a function of the loss on ignition as a percentage of the dry weight (% dry wt) So long as heavy metals are preferentially bound to organic particles (and to the fine particulate fraction) there must be a relationship between the loss on ignition and, e.g., the zinc content of the samples — expressed in Fig. 14.8 in parts per thousand of the residue in ignition. A proviso is a restricted catchment area. In Fig. 14.8 the values relate to solids form the Neckar between Stuttgart and Poppenweiler obtained between 1970 and 1973. The limitations in respect of time and space in the preparation of correlations must be noted.

That such relationships become weaker for catchments of greater extent can be seen from the analytical results for Elbe sediments with respect to polychlorinated biphenyls, for which concentrations are plotted against the loss on ignition of the

Zinc (°/oo ash on ignition)

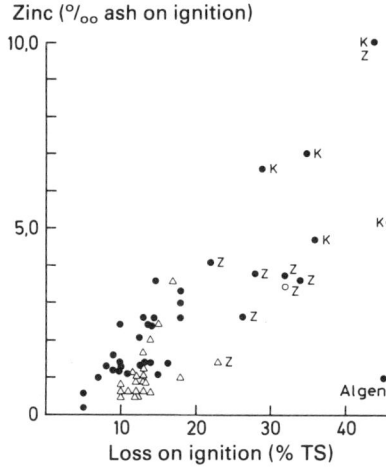

Fig. 14.8 — Relationship between loss on ignition and Zn content for suspended solids samples in the middle Neckar (Stuttgart region), 1970–1973. Key: K = sewage effluent; Z = centrifuge samples; △ = 1970 values.

samples (Fig. 14.9). Doubtless the measurements would show a much closer correlation within a more confined area, such as a dock basin, for example. In addition, the tracer function of phosphate is demonstrated in Tables 7.2, 8.2, 9.3 and 9.5.

Correlations between one tracer and certain trace substances, such as have already been discussed for sludges, are also demonstrable with reference to the analysis of recent, non-polluted or scarcely polluted sediment deposits. The same applies to suspended solids in flowing waters. Fig. 14.10 gives an impression of the type of correlation existing between loss on ignition and the level of hydrocarbons for sediments and suspended solids of different origin.

In the technical literature the idea is sometimes advanced (Symader *et al.*, 1977) that a particular correlation, once established, can be used in order to dispense with the need for direct determination in respect of the more difficult analytical parameters, as in the statement 'With such a clear dependence it becomes superfluous to measure both PO_4 and also gamma-HCH'. A responsible analyst can only subscribe to this opinion when it is clear that there is a causal relationship between the tracer parameter and the trace substance in question.

14.5 ASSOCIATED PARAMETERS

In the course of an intensive study of analytical data for samples of different origin in time but always from the same place, such as, for example, fortnightly suspended solids samples, one may sometimes observe a striking parallelism. Thus the contents of different heavy metals may be correlated with each other, as may be seen for lead and zinc in Fig. 14.11, and the phosphate P may perhaps be included in the correlation as well. As this mainly concerns sediments from a wide catchment area,

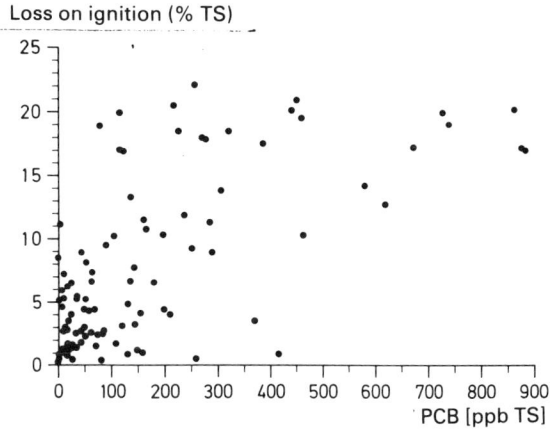

Fig. 14.9 — Relationship between loss on ignition and PCB content of sediments from ARGE-Elbe 1983.

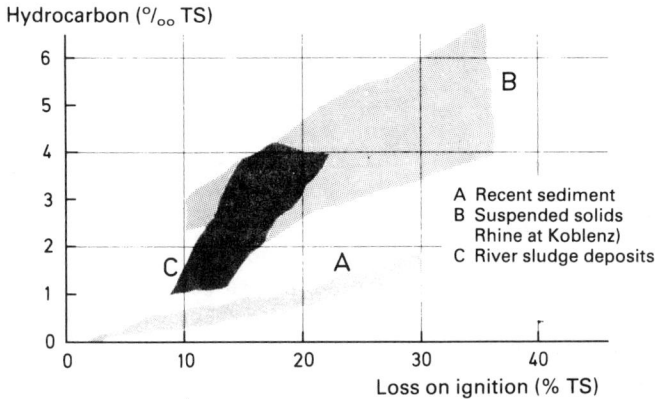

Fig. 14.10 — Relationship between loss on ignition and hydrocarbon content of suspended solids, sediments and sludges.

this observation can hardly be considered under the heading of 'identical origin', such as has already been referred to in Section 14.3. Very many point source discharges are certainly involved in determining the pollution level of the suspended solids, which in their totality counterbalance the smaller irregularities. Under these circumstances the similarly shaped profiles for different determinands should not be thought remarkable, but rather the deviation of, say, one particular heavy metal from the general trend. Then the analysis should be checked, and should the initial

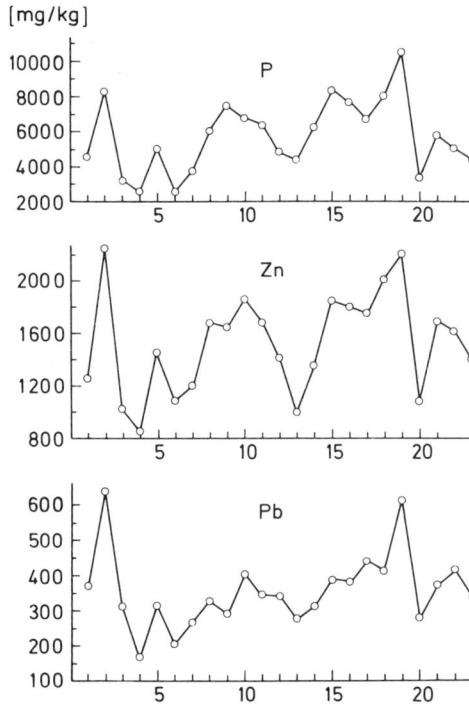

Fig. 14.11 — Common patterns pf Phosphate P, Zn, and Pb contents in suspended solids from
the Rhine at Bimmen/Lobith, January–December 1979.

results be confirmed the particular cause of such behaviour must be sought by
searching for a special point source.

Not only the suspended solids samples obtained by settling from the Lower
Rhine, but also the combined 14-day samples obtained in an automatic sample
collector, may contain, besides lead, substances like nickel and arsenic which tend to
follow the lead contents up and down. This is by no means understandable from an
overall determination of dissolved and undissolved substances, as both fractions
behave differently in response to changes in the discharge regime (Fig. 7.4 *et al.*). For
an exact evaluation one would then find that — despite a generally similar trend —
there is not a constant relationship between the two parameters, as can be seen from
the pair lead/arsenic in Fig. 14.12.

In closing, an exception in the case of the level of contaminants in fish flesh should
be briefly mentioned. According to Fig. 14.13 hexachlorobenzene contents in fat
from the muscle of Elbe bream were broadly correlated with the gamma-HCH
contents. It is at first of secondary importance which of these is regarded as the
associated parameter. More important is the basically similar behaviour of the two
organochlorine compounds, which both exhibit the same bioaccumulation tendency.
By a more precise identification with the 'matrix' (fat), the relationship between the
two determinands becomes even closer, as was apparent from Fig. 14.9 for the loss
on ignition and the PCB content of the entire sample.

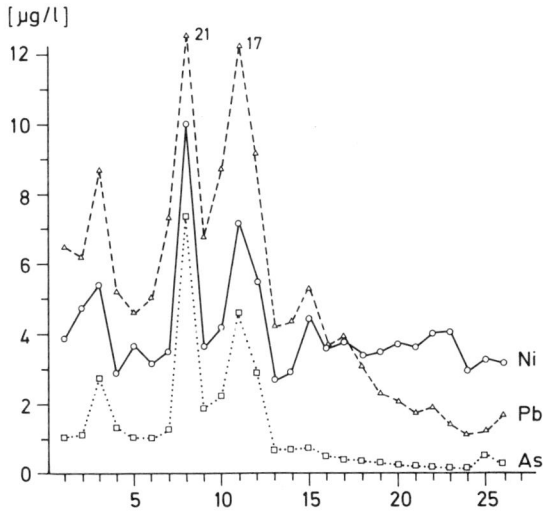

Fig. 14.12 — Common patterns for As, Ni and Pb contents in suspended solids from the Moselle at Koblenz in 1983.

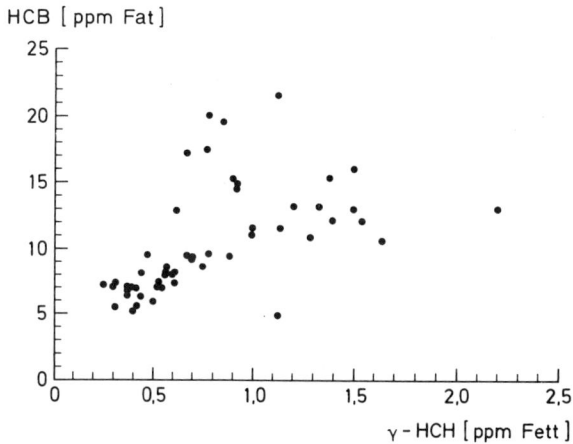

Fig. 14.13 — Relationship between γ-HCH (Lindane) and HCB-contents in fat from Elbe bream, from ARGE-Elbe 1983.

LITERATURE

14.2

Arbeitsgruppe des VCI-AWA-Arbeitskreises "Analysengeräte" (1977). Untersuchungen zur Vergleichbarkeit des chemischen Sauerstoffbedarfs (CSB) und des gesamten organischen Kohlenstoffs (TOC), wlb "wasser, luft und betrieb" 21, 182–183.

Kempf, T. (1970). Die Bedeutung der Bestimmung organischer Stoffe im Wasser, Schriftenr. Ver. Wasser–Boden–Lufthyg. Berlin–Dahlem, H. 33, S. 105–115.

Sontheimer, H. (1970). Untersuchungen zur Belastung des Rheins mit organischen Stoffen. gwf-Wasser/Abwasser 111, 420–426.

Hauptausschuß für Meß- und Kontrolleinrichtungen der Wassergütewirtschaft (1975). Instrumentelle Bestimmung der organishen Stoffe in Wässern, Z. Wasserund Abwasser-Forsch. 8, 164–175.

Winkler, F., Düwert, B., Patecki, H. (1984). Möglichkeiten der Korrelation unterschiedlicher Wasseranalysenverfahren. Acta Hydrochim. Hydrobiol. 12, 239–246.

Sontheimer, H., Kühn, W. (1981). Das spektrale Absorptionsmaß bei 254 nm als Anagabe für die Konzentration an Störstoffen, DVGM-Schriftenreihe–Wasser Nr. 26, Frankfurt a. Main (Kap, 6.1).

14.4

Symader, W., Herrmann, R., Rump, H.-H., Richartz, H. (1977) Der Summenparameter und ein anderer Weg zur Vereinfachung der Gewässergüteüberwachung, Dtsch. Gewässerkd. Mitt. 21, 118–120.

Hellmann, H., Griffatong, A. (1972). Herkunft der Sinkstoffablagerungen in Gewässern. 1. Mitt., Chemische Untersuchungen der Schwermetalle, Dtsch, Gewässerkd. Mitt. 16, 14–18.

14.5

Arbeitsgemeinschaft für die Reinhaltung der Elbe (1983). Chlorierte Kohlenwasserstoffe – Daten der Elbe, – Hamburg.

15

Calculating rates of mass transport

15.1 INTRODUCTION

Some analysts are of the opinion that it is impossible to determine rates of transport of water constituents with any degree of accuracy. It is a fact that the quantity which is transported each second or in a longer period through the entire flow cross-section can only be measured approximately, for a number of reasons. Apart from the problems of representative sampling and analytical accuracy, there is also the question of the homogeneity of the water body, as regards both the geometric cross-section and also its variation in time. Certainly it is a well-known fact that the suspended solids content of flowing waters increases with distance below the surface (Table 5.1) and may also exhibit considerable diurnal variation (Fig. 5.1). This chiefly affects all those substances which are bound to the suspended solids fraction. It may further be remarked that the analytically measurable concentration of a given substance is only applicable to mass transport calculations in conjunction with the corresponding value of the discharge. *De facto*, relatively few gauging points for discharge measurements are present along German rivers. (These are listed in the German Hydrological Yearbooks for the Federal and Provincial Offices of Water Management.) Where samples are taken at points *in between* two gauging stations, approximate values for the discharge must suffice. Even for the Rhine at Koblenz, the rate of flow through outlets from the Rhine at Kaub and from the Lahn at Kalkofen must be calculated. The accuracy of the daily mean for most gauges is quoted as ±1.5%, but usually one has to be content with an error of ±3%.

Much more serious than these obstacles to mass transport calculations, however, are the changing concentration values with rising and falling values of the discharge. In the strict sense there is no closely defined water body in which the concentrations of *all* parameters are reproducible along the time axis, hence the instantaneous situation at the time of sampling is not reproducible. For the interpretation of annual mass transport values there is yet another major problem: the relevant annual discharges differ widely from year to year (Table 15.1) and hence, as will be shown, the annual mass transport values differ likewise.

This means that the mass transport for a flowing water body cannot be exactly

Table 15.1 — Mean annual discharges for the Rhine (Kaub) and Moselle (Cochem)[a]

Year	Rhine (m^3/s)	Moselle (m^3/s)
1960	1350	204
1961	1670	334
1962	1560	335
1963	1280	134
1964	1100	202
1965	2010	352
1966	2260	460
1967	1920	366
1968	2060	443
1969	1700	331
1970	2230	462
1971	1180	186
1972	984	154
1973	1380	196
1974	1440	241
1975	1920	330
1976	1080	138
1977	1690	283
1978	1950	387
1979	1700	380
1980	2010	423
Average	1642	302

[a]Source: Federal Office of Water Management.

quantified for either a second, a month, or even a year. All the same, mass transport data are indispensable for numerous purposes, even though they can only be estimated within a margin of 10 or 20%. Such data are absolutely necessary for mass balance calculations in respect of substances entering the environment, and especially from the standpoint of their persistence. The data are just as essential for determination of trends, and a further aspect not to be overlooked is the ecologically and politically important topic of the pollution of coastal waters by inland waterways.

As an exception to the general rule it is possible, at the cost of considerable time and effort, to obtain transport data for a period of a day or a few days (e.g., in connection with an acutely increased level of contamination of a watercourse—see Section 13.5) to a fair degree of accuracy. More difficulties are associated with the annual mass transport rates. Data occasionally reported in the literature which are based on the scale-up of only a few results without any reference to the discharge relationships prevailing at the time certainly do not comply with the conditions which must be observed for a serious evaluation; apart from any other considerations, any accompanying estimates of trends must be considered very questionable.

A better method is always more expensive! Even for a rough scale-up designed to give an indication of the order of magnitude, properly verified results are a necessity. The best assurance and probably the only really acceptable method is to use transport/discharge relationships calculated for gauging points on the river in question.

15.2 TRANSPORT/DISCHARGE RELATIONSHIP FOR DISSOLVED SOLIDS

Transport/discharge relationships were referred to in several sections in Part II. According to data presented in Chapter 6, the chloride transport rate in the Rhine at Stein increases linearly with the discharge (Figs 6.3 and 6.5), and similarly the sulphate transport (Fig. 6.6). The correlation in each case is really close. The nitrate transport rate in the Rhine at Koblenz/Kaub similarly increases in a linear fashion with the discharge (Fig. 6.9), although with reservations in respect of the scatter of individual values, particularly at high discharge values, which give the relationship a funnel shape. At this point no more can be said about the causes of this variation (e.g. as a consequence of water temperature) and the source of the substances concerned, although much might be deduced from the graphs.

One may, however, state the fact that even for equal values of discharge the transport rates are subject to certain fluctuations, the more so the smaller the watercourse in relation to the catchment area or the size of the 'sewered' population (Fig. 4.1 and Table 4.4). An increase in the mass transport rate with discharge is therefore a very widespread phenomenon, which can even be observed for substances for which it may not *a priori* have been expected (Fig. 11.8).

Within this general state of affairs, a strictly linear transport/discharge profile becomes more or less the exception. Where it does obtain, then of course the annual mass transport or the annual mean transport rate can in fact be quoted as a rate per second, as long as the analysis supports this. While for chloride and sulphate at Stein am Rhein the results do in fact support this, nevertheless the chloride mass transport rates for the Middle and Lower Rhine present the opposite extreme (Fig. 15.1). The daily transport rates in this case vary so widely that their correlation with the discharge is of little or no meaning. In the practice of surface water analysis one has to take into consideration whether the chloride content is determined daily or, as in some places, continuously.

Other parameters may not be so easily or frequently determined as chloride. If one takes as a typical example of a trace substance the dissolved zinc in the Rhine at Koblenz 1977–79, then here too there is no gainsaying an increase in transport rate with discharge. Besides the noticeable scatter of the individual values even for approximately equal discharge values, which may reach 100%, there are some particularly high transport rates with the character of 'outliers'. With the incursion of statistical concepts and modes of interpretation into the data evaluation process, outliers of this kind have sometimes been wrongly eliminated. It is, however, correct that one can obtain the data required with the desired accuracy, by totting up transport rates determined daily. What is more, certain alternative approaches have developed by reason of the personal effort called for. Thus some authors first of all use the collected data to calculate a smoothed curve. From this they deduce the value corresponding to a particular discharge, such as the MQ value. If one assumes for the

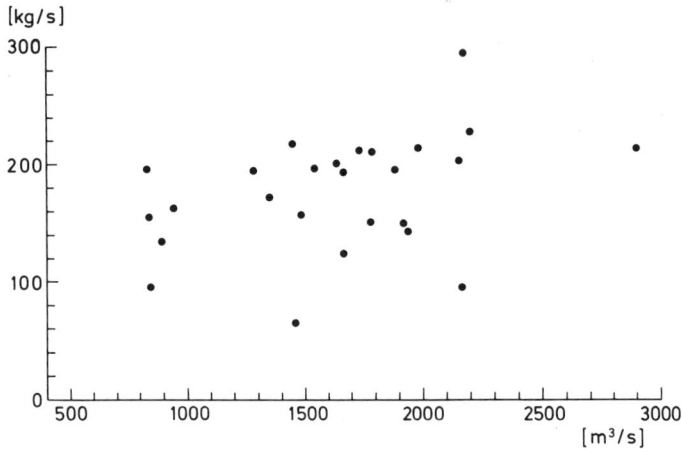

Fig. 15.1 — Relationship between chloride transport and discharge in the Rhine at Koblenz, 1977.

case of dissolved zinc an average discharge of 1560 m^3/s, then the corresponding zinc transport rate can be calculated with *mathematical* accuracy together with upper and lower limits (the term 'standard deviation' in this connection is out of place). For all estimates using the auxiliary apparatus of statistics, one must not overlook the fact that under primarily hydrological circumstances the analytical results include the 'coordinates' of the sampling place and time. Even if the relation between discharge and transport rate is very close it cannot be regarded as a 'function'. To this extent the smoothed curve cannot be equated with the type of curve derived from pairs of values of purely analytical significance, such as those obtained during the calibration of equipment or from controlled ring tests. And they are also different from the analogue relations employed in the context of sewage treatment.

From visual inspection of Fig. 15.2 a value for the average rate of mass transport of 34 g/s is obtained (MQ=1560 m^3/s from the 1936–65 annual data†). The lower limit of about 25 g/s is equivalent to 788 t/annum and the upper limit (45 g/s) to 1410 t/annum. The analyst may therefore quote a value of 1100 t/annum for the mass transport of zinc with a possible error of ±50%, as long as extreme discharge values are not involved.

In so far as the few values for the years 1977 and 1978 in Fig. 15.2 permit any indication, the annual tranport figures did not alter to any measurable extent during the period from 1977 to 1979.

A further informative example is given by the level of dissolved cadmium, for which the concentration varied, according to the IKSR Data Books, from <0.1 to 0.2 µg/l in the Middle Rhine during 1979 and hence was about two orders of magnitude below that of zinc (Table 15.2).

It is surprising that the cadmium concentration is evidently independent of discharge. From hydrological considerations, this would only be possible if the

† The mean for the period 1970–80 from Table 15.1 is slightly higher (1642 m^3/s).

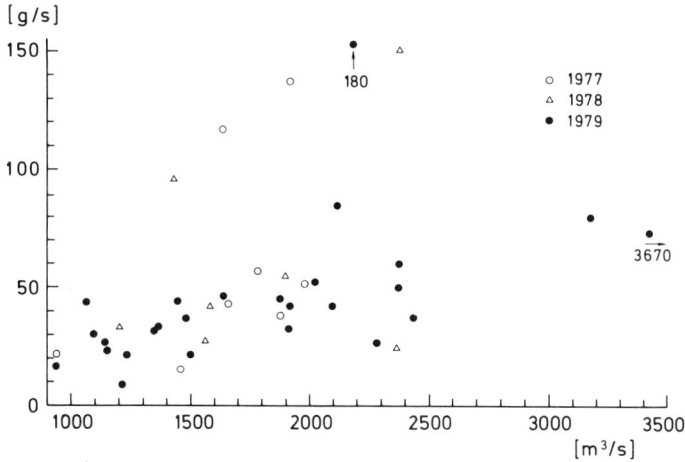

Fig. 15.2 — Relationship between dissolved Zn transport and discharge in the Rhine at Koblenz.

cadmium was derived exclusively from natural sources, i.e. the background. As this is hardly feasible, a further possible explanation lies in the analysis itself. For the present example, this is obviously incapable of lowering the measurement limit in the prevailing concentration range and hence of differentiating sufficiently between existing values. The practising analyst is of course aware of how difficult it is to avoid contamination, even where cadmium is concerned, and how great the risk of contamination must be in a 'normal' laboratory.

If despite these difficulties one bases calculations on a plausible value of $0.1~\mu g/l$, then for $MQ = 1560~m^3/s$ one obtains an annual transport of 4.9 t. At the Dutch–German frontier, concentrations were appreciably greater (Fig. 15.3) and in addition a hyperbolic decrease with increasing discharge is apparent. If one disregards the considerable fluctuations in the concentration for equal water flowrates, and assumes an average of $0.6~\mu g/l$ for $MQ = 2200~m^3/s$ (average for the years 1936–1965), then the mass of cadmium transported by the Rhine in 1979 amounted to 44 tonnes.

In closing this section, mention may also be made of the approximately linear transport/discharge relationship for humic acids in the Rhine (Fig. 13.7).

15.3 TRANSPORT/DISCHARGE RELATIONSHIP FOR UNDISSOLVED SOLIDS

The undissolved (trace) substances are bound to the suspended solids. For the latter the concentration increases exponentially with the discharge (Fig. 2.1). Thus for the suspended solids transport there is also a typically steep exponential increase with discharge. As, however, numerous results, including those in Fig. 7.9, demonstrate that the specific pollution level of the suspended solids decreases with increasing concentration on or mass transport of suspended solids, it remains to be shown how the trace substance flux behaves in the same situation. If one starts with the

Table 15.2 — Discharge values and concentrations of dissolved cadmium in the Rhine at Koblenz in 1979 (data from IKSR Tables)

Discharge (m^3/s)	Cadmium (μg/l)	Date (1979)
928	<0.1	20 September
1060	<0.1	24 January
1100	0.1	3 October
1140	0.2	2 November
1150	0.1	7 August
1210	0.1	29 July
1240	0.2	21 October
1350	0.1	3 September
1360	0.1	20 August
1450	0.2	10 December
1480	<0.1	10 January
1500	0.1	13 July
1630	0.1	1 December
1910	<0.1	21 April
1910	0.1	28 December
2020	0.1	31 May
2100	<0.1	24 February
2120	<0.1	9 March
2190	0.1	12 June
2290	<0.1	25 June
2380	0.2	7 April
2380	0.1	6 May
2440	0.1	18 November
3180	<0.1	6 February
3670	<0.1	19 March

hypothesis that the *emitted* mass flow is constant in time, then the result must necessarily be a constant mass transport rate in the watercourse.

On the contrary the transport of undissolved trace substances almost always increases with discharge. Often the transport/discharge relationship is certainly very loose as, for example, for undissolved zinc in the Middle Rhine in 1979 (Fig. 15.4). The values applicable to a given discharge vary much more widely for high values of discharge than they do for lower ones. This is doubtless attributable to the special circumstances prevailing during rising and falling flood flows. That the measurements assume a pear-shaped profile in the graph under such conditions has already been shown (Fig. 6.14). Of course the assumption of a pear-shaped profile does not make the estimation of the associated annual transport rates any easier! Nevertheless the transport/discharge relationship, however unsatisfactory it may be, has to suffice

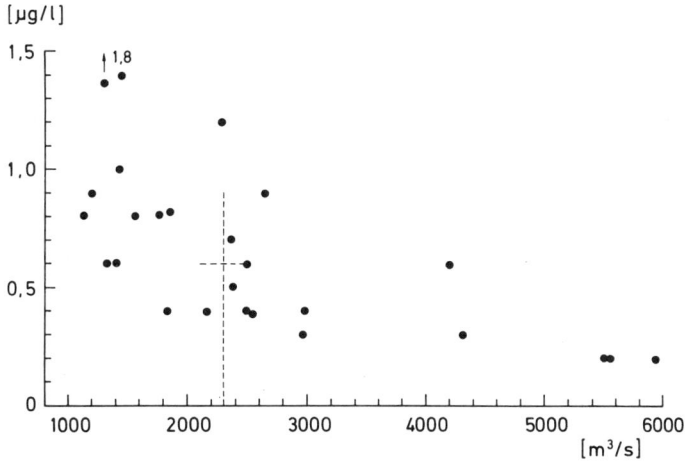

Fig. 15.3 — Relationship between concentration of dissolved Cd and discharge in the Rhine at Bimmen/Lobith, 1979.

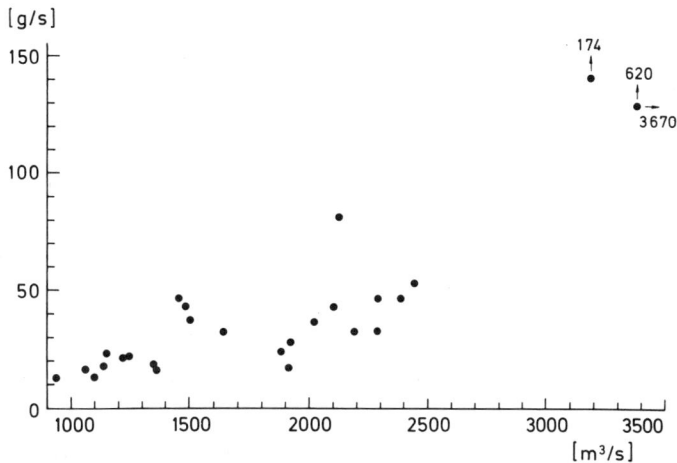

Fig. 15.4 — Relationship between the mass transport of undissolved Zn and discharge in the Rhine at Koblenz, 1979.

for calculation of the annual mass transport. According to Fig. 15.4 a rate of 30 g/s may be derived for the MQ of 1560 m³/s. The annual transport of zinc thereby works out at 950 t, which approaches that for dissolved zinc. The actual quantity of zinc transport by the water, depending on the homogeneity of the water body in the stream cross-section and assuming accuracy of the reported data, may in fact be 50% higher or lower according to the hydrological position, especially the discharge behaviour, for otherwise equal inputs. While this may appear to be illogical in the case of 'equal inputs', the following factors must be introduced into the argument: high discharges imply enhanced erosion in the catchment area (background!), enhanced entrainment of previously deposited contaminants, and poorer treatment and separation efficiency within the ambit of sewage treatment works, the latter especially during periods of elevated rainfall and stormwater runoff. Concerning the construction of transport mass balances, more will be said in Chapter 17.

We now come to an example from the range of trace organic impurities. The values of mass transport for undissolved hydrocarbons in the Middle Rhine for 1983/84, determined on the basis of spot samples, illustrate the state of affairs that has already been mentioned several times. Thus while on the one hand the specific pollution level of the suspended solids (mg/kg) falls with increase in the discharge and suspended solids transport, and on the other hand the increase in suspended solids content simultaneously involves an increase in the hydrocarbon concentration in the water volume (mg/l), there is an exponential increase in the latter quantity (Fig. 15.5). While for a linear (funnel-shaped) transport/discharge relationship the

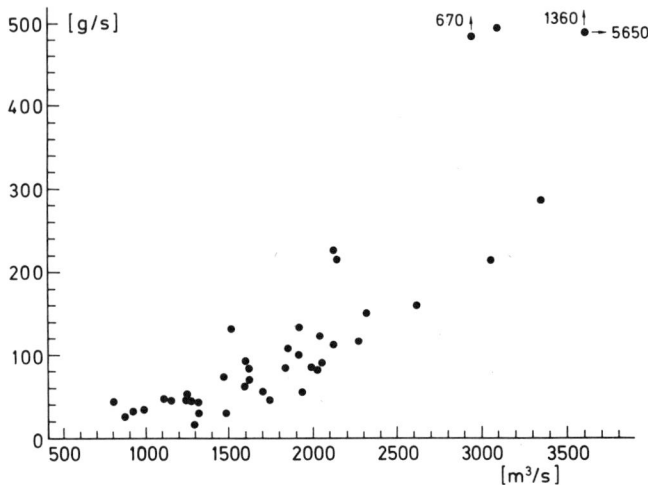

Fig. 15.5 — Relationship between the mass transport of hydrocarbons and the discharge in the Rhine at Koblenz, 1979.

annual mass transport can be scaled up from the average discharge (MQ) in m^3/s and the corresponding transport rate, even if with certain limitations, with an exponential profile this is beset with even greater uncertainties. The 40 pairs of figures obtained between 4 November 1983 and 15 August 1984 (Fig. 15.5), for example, give an average mass flowrate of 128 g/s. The corresponding value derived from Fig. 15.5 on the basis of the MQ is, however, only 68 g/s. Accordingly, after scaling-up to annual transport figures, widely differing amounts of 2140 and about 4000 t/annum are obtained.

Another kind of 'scale-up' starts from the specific pollution level of the suspended solids 'normalised with respect to MQ' (2000 mg/kg) and the relevant analytically determined suspended solids content at Koblenz (25 mg/l). This calculation gives a figure of 2460 t/annum.

There is certainly little sense in departing from hydrologically well-founded values when dealing with the scale-up calculation. On the other hand, for many practical purposes only the mass transport values are relevant, and then the unavoidable errors, as long as they are estimated and quoted, are frequently tolerated. This will become apparent when some practical examples are considered in Chapter 17.

15.4 TRANSPORT/DISCHARGE RELATIONSHIP FOR DISSOLVED/ UNDISSOLVED SOLIDS

Numerous water constituents occur in both dissolved and undissolved states. The transport/discharge relationship can therefore be related to the sum of dissolved+ undissolved fractions. Substantially more information is, however, gained from the separate relationships for the dissolved and undissolved fractions, which even if not exactly counterpoised nevertheless exhibit different slopes (Fig. 15.6). From the

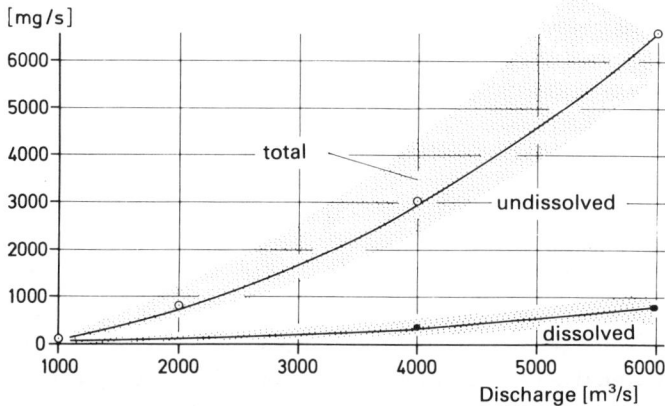

Fig. 15.6 — Relationship between the dissolved and undissolved PAH fractions and the discharge in the Rhine at Bimmen/Lobith, 1977–79.

hydrological viewpoint this is also the only acceptable method of evaluation. With sufficient data the relationship expressed in Fig. 15.6 may be somewhat standardised, in that the scatter of the individual values is covered by the band (shown dotted). Regarding the PAH transport rates indicated, the graph supplies the following information. The dissolved fraction increases steadily with discharge; in the range of low to medium flows, MQ=2200 m³/s, the transport rate is practically constant and amounts to 150 mg/s which is equivalent to 0.065 μg/l—for more details concerning PAHs see Chapter 9. The undissolved fraction is thus many times greater, and is closely connected, via the water flowrate, with the rate of suspended solids transport. The annual transport corresponding to the MQ is around 30 t. A noticeable feature is the steep increase in the region of flood flows.

More difficult or more ambiguous questions are involved in evaluating those results which, as for the phosphate ion, deal with three different, chemically distinguishable fractions, all of which may interact with each other. By this we imply the orthophosphate, the undissolved phosphate and the hydrolysable dissolved phosphate. The hydrolysable fraction consists largely of the so-called polyphosphates and the organic combined phosphates. The values for each of these fractions vary both in magnitude and also in their relative proportions, depending on the sampling point (agricultural usage, degree of urbanisation, etc., in the catchment area) and the discharge values. Analytically one proceeds so as to determine separately the total phosphate, the suspended solids-associated undissolved phosphate and the orthophosphate. On the basis of numerous measurements, and with appropriate smoothing of the respective curves, transport/discharge relationships can be constructed which for polluted rivers exhibit a pattern resembling that of Fig. 6.15, whereby the transport of hydrolysable dissolved phosphate increases most rapidly with the discharge, followed by the transport of undissolved phosphate.

In a particular case the respective concentrations may deviate considerably from the profile represented by the smooth curve, as is very apparent from Fig. 6.14..

The foregoing discussion leaves out one very important 'dimension', namely the interaction of dissolved with undissolved fractions, or in other words the distribution equilibrium between water and suspended solids components. According to information already to hand, the transport/discharge relationship can provide useful clues to this aspect. However, this particular sub-topic is so extensive and at present so little understood that the very little information there is would probably be more confusing than helpful.

15.5 CONCLUSIONS

The transport/discharge relationship may exhibit one of three forms (Fig. 15.7). The almost linear Curve A is encountered chiefly where dissolved substances of natural origin are supplied from the catchment area. They are directly associated with erosion phenomena. More often the picture is of a roughly linear relationship, but with an increasingly wide range of scatter of individual values (B). A transition to a more exponential form of transport/discharge relationship (C) occurs continuously, which is not solely attributable to the high degree of scatter in the region of elevated discharge.

The linear relationship allows the annual mass transport for a particular substance to be predicted quite accurately, based on a knowledge of the appropriate annual total flow. For practically all constituents of anthropogenic origin, however, this cannot be done. Thus the question of an alternative approach must be faced. On the basis of experience a 'representative value' is adopted for the purpose (Sontheimer). As it is apparently ideally suited for this purpose, the average discharge (MQ)—transport value is selected where the Rhine is concerned. As it is also

Mass transport

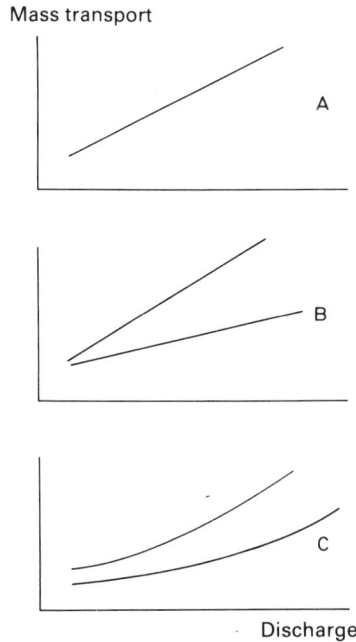

Discharge

Fig. 15.7 — Mass transport/discharge relationships obtaining in flowing watercourses.

possible that different mass transport rates may occur even for average discharge values, then once again an average value must be selected. By the use of representative values obtained in this way, the daily and annual mass transport estimates are obtained, which are as far as possible independent of chance fluctuations in the annual discharge (Table 15.1). The great advantage of this method lies in the comparability of annual mass transport figures. As long as one refrains, for very good reasons, from using this method for such specialised purposes as the calculation of the input of contaminants to inshore waters during a limited period, then there remains only the tedious estimation of the daily transport rates.

Even if, for the Rhine, the mass transport associated with the average rate of discharge is taken as the 'representative value', nevertheless for other rivers this approach based on the MQ value appears definitely less reliable. A glance at Table 15.3 will show that from one river to another, different proportions of flood flows are included in the MQ value. Explained in pictorial terms this is expressed by a wide-mouth funnel-shaped profile for the transport/discharge relationship. This state of affairs does not help the quality of the scaled-up mass transport data. A general solution to this problem must await the outcome of further investigations.

The representative value stands or falls with the degree of suitability of the sampling point. A first requirement is a guarantee of homogeneity of the water body over the channel cross-section. This condition is usually fulfilled in the smaller, free-flowing rivers. Exceptionally difficult problems are presented, however, by slow-moving, regulated reaches, and also by large, wide watercourses shortly below

Table 3 — Discharges of various watercourses, average values from the 1936–65 annual data[a]

River and gauging point	MNQ (m^3/s)	MQ (m^3/s)	MHQ (m^3/s)	$\dfrac{MQ}{MNQ}$	$\dfrac{HQ}{MQ}$
Rhine/Kaub	702	1560	3950	2.2	2.5
Neckar/Rockenau	28.3	121	1118	4.3	9.2
Main/Kleinheubach	37.1	153	686	4.1	4.5
Moselle/Cochem	58.9	288	1900	4.9	6.6
Ruhr/Wetter	9.72	65.4	765	6.7	11.7
Lippe/Schermbeck	14.8	49.5	290	3.3	5.8

[a]Data from German Hydrological Yearbook, Rhine catchment, 1968 issue.

massive effluent outfalls. In the case of spot samples, some information must also be available about the diurnal fluctuation of the concentration of the substance, and also the possible effect of water temperature on the results. Of fundamental importance also is clear knowledge on the part of the analyst of the probable range of error of his determinations.

LITERATURE

15.2—15.4

Schroeder, G. (1950). Die Korrelationsrechnung und ihre Anwendung in der Wasserwirtschaft, Hrsg. Bundesanstalt für Gewässerkunde, Bielefeld.

Doerffel, K. (1962). Beurteilung von Analysenverfahren und -ergebnissen, Springer-Verlag, Berlin.

Sontheimer, H., Gimbel, R., Weindel, W. (1980). Die Rheinwasserqualität – Vorschlag einer neuen Darstellung aus der Sicht der Trinkwasserversorgung, IAWR 7, Arbeitstagung 1979, Amsterdam.

Pöpel, J. (1971). Schwankungen von Kläranlagenabläufen und ihre Folgen für Grenzwerte und Gewässerschutz, Schriftenreihe Wasser-Abwasser, H. 16, München–Wien.

Hellmann, H. (1974). Zur Inhomogenität von Flußwasserparametern, Dtsch. Gewässerkd. Mitt. **18**, 23.

Wagner, G. (1972). Die Berechnung von Frachten gelöster Phosphor- und Stickstoffverbindungen aus Konzentrationsmessungen in Bodenseezuflüssen, Bericht Intern. Gewässerschutzkommission für den Bodensee.

Internationale Kommission zum Schutz des Rheins gegen Verunreinigung, Zahlentafeln für jedes Kalenderjahr.

Internationale Kommission zum Schutz des Rheins gegen Verunreinigung, Langfristige Arbeitsprogramme (LZP) sowie Tätigkeitsberichte (z. B. 1978–1983 jährlich), Koblenz.

Süßmann, W. (1985). Vergleich der Belastung zweier Bäche im ländlichen Raum anhand mehrjähriger Untersuchungen. Z. Wasser-Abwasser-Forsch. **18**, 56–66.

Sontheimer, H., Giebel, R. (1977). Untersuchungen zur Veränderung der Fracht an organischen Wasserinhaltsstoffen mit der Wasserführung am Beispiel des Rheins, gwf-Wasser/Abwasser **118**, 165–173.

Zobrist, J., Davis, J. S., Hegi, H.-R. (1977). Charakterisierung des chemischen Zustandes von Fließgewässern, Gas-Wasser-Abwasser **57**, 402–415.

Hellmann, H. (1968). Die Belastung des Rheins mit ungelösten organischen Stoffen im Jahre 1967, Dtsch. Gewässerkd. Mitt. **12**, 39–43.

Hellmann, H. (1970). Untersuchungen zur Kohlenwasserstoff-Fracht des Rheins 1968/69 und Überlegungen zu deren Herkunft, Dtsch. Gewässerkd. Mitt. **14**, 14–18.

Hellmann, H. (1985). Verhalten von leichtflüchtigen Chlorkohlenwasserstoffen in Fließgewässern, Z. Wasser-Abwasser-Forsch. **18**, 210-216.

16

Trend determination

16.1 PROBLEMS OF TREND DETERMINATION

In the context of surface water analysis, the term 'trend' implies the progressive change in concentration or mass transport rate for a particular substance with the passage of time. The study of this phenomenon, which may reveal an increasing or decreasing trend, or even a static situation, is one of the main objects of analytical work, as the trend is a demonstration of the effects of changing economic, hydrological and possibly legislative measures on water quality over a period of time. Well-known examples from the recent past include the efforts put into the reduction of degradable organic matter, heavy metals and phosphates.

If one surveys the changes occurring during the last two decades in this respect, the recognition of a particular trend will most probably be feasible only given proper analytical techniques, optimal choice of sampling station, a sufficient number of results, and due regard to the oft-mentioned hydrological conditions.

The first-named condition is by no means fulfilled in all water laboratories. In a ring test which was concluded in mid-1985, for example, one water sample which had been spiked with 3 μg Cr/l was analysed by 13 participating laboratories. Deviations of between 1 and 12 μg/l were obtained in the measured value and even the best groups might show a standard deviation of 15% or more in their results. It is obvious, therefore, that a trend in a positive or negative direction can only be detected where the range of error in the analytical results is lower than the changes to be expected in the parameter under investigation. A uniformly occurring systematic error with otherwise acceptable reproducibility, however, only interferes slightly with the recognition of a trend. It will be seen that in the past many parameters were apparently 'amenable' to the detection of trends, and by implication lent themselves to routine analysis, while others placed much greater demands on the analytical resources and called for meticulous performance by the analyst.

One further point may be mentioned by way of introduction: that of trend determination for sediments. Experience shows that numerous parameters are more readily determined in sediments than in the water phase as they are present in much higher concentration, which for some trace substances may exceed that in the water phase by as much as 1000-fold. The problem of sediment analysis then consists much more in the use of the correct method of sampling from undisturbed deposits, and, as long as sediments were laid down without interference due to extraneous forces, in the dating of the successive segments from a vertical sediment profile. Sludge deposits are, with few exceptions, unsuited for the purpose of trend estimation.

16.2 FRESHWATER LAKES

The literature contains numerous examples of the progressive increase of phosphate concentrations in freshwater lakes. In the case of Lake Constance and the orthophosphate content, a relatively slow rise was apparent up to 1958, which then became steeper between 1958 and 1969 and was even more pronounced after 1970 (Fig. 16.1). Parallels have also become apparent among the Swiss lakes. The trend profile

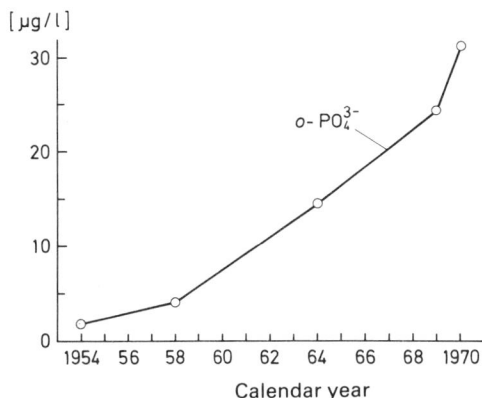

Fig. 16.1 — Development of o-PO$_4^{3-}$ concentration in Lake Constance; values determined on completion of turnover (from Bernhardt, 1972).

strongly resembles that for household detergent consumption, and less closely reflects the trends in phosphate application in agriculture.

A corresponding trend with regard to phosphate and nitrate contamination is also discernible from the sediment profiles for Lake Constance, provided the sediment cores were obtained from appropriate locations (Muller 1977). The particular conditions for such trend estimations are complied with more especially in the deepest part of that lake, as already mentioned in Chapter 7 (Fig. 7.14). The different results also permit the calculation of annual phosphorus inputs to the lake. The scaled-up estimates confirm the more or less exponential rise from 1958 onwards, from a phosphorus figure of 600 t/annum to almost 2000 t/annum a few years ago (Wagner 1976).

Trace metals such as cadmium, lead and zinc as well as the trace organic contaminants benz(a)pyrene, DDT and lindane and the hydrocarbons in general largely conform to the same pattern as phosphate and show definite indications of a trend (Muller 1983).

The author's own results for Lake Constance sediment profiles (Table 8.2) provide additional confirmation of these extensive results. The sediment core referred to in Chapter 10 (Table 10.4) from the Lower Ems at Diele is also useful in connection with trend estimation on account of the conditions under which it was believed to have been formed. The results did not, however, provide any evidence of a rise or fall in the specific level of contamination, depsite the period of sedimentation, covering roughly three decades. Other parallel cases are known in which a constant level of sediment pollution, chiefly with PCBs, is exhibited by the vertical sediment profile.

16.3 LIGHTLY POLLUTED WATERCOURSES

The hydrological conditions under which it is possible to study the trends for water constituents will be explained more fully in the ensuing section. In principle the method of trend determination is independent of the level of pollution of the water body concerned. If therefore a special section is devoted to lightly polluted waters, these may be regarded as a special case, more readily understood.

If one looks first at the presence of the classical parameters in the Upper Rhine (Chapter 6) then it is possible to discern a rising trend in the chloride content between 1965 and 1975 from a glance at the transport/discharge relationship (Section 6.5). If one wishes to define this trend more precisely, then it is necessary, according to the procedure of Chapter 15, to estimate the annual mass transport for chloride for a particular discharge, which must be specified in more detail, and to compare the resulting figures for successive years. In *this* case it seems unimportant *which* value of discharge is selected, as the transport/discharge profile is practically linear. For sulphate (Fig. 6.6), no significant change in the linear correlation was apparent over the 10-year period.

The concentration of orthophosphate ions in the Upper Rhine and at the sampling station of Stein am Rhine (1 million population in the catchment area, $MQ=300–400$ m^3/s) rose from 0.02 to 0.11 mg/l between 1961 and 1975 (Reichenberger, 1976). The graph (Fig. 16.2) presents mean values for the water samples

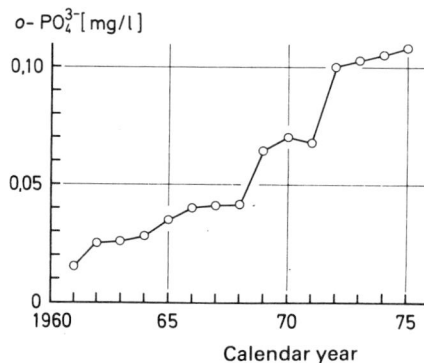

Fig. 16.2 — Development of o-PO$_4^{3-}$ concentration in the Rhine at Stein (from Reichenberger (1976) and IKSR Data).

obtained from April to August in each year. The (disregarded) fluctuations in discharge have not concealed the very noticeable trend observable at this point, as would have certainly occurred in the case of other sampling stations further downstream. In other respects the trend profile closely resembles that for Lake Constance water illustrated in Fig. 16.1.

Climatic conditions can often influence the results, even for roughly similar values of discharge, a fact which can be seen from the values of the classical parameters obtained by analysis in spring and autumn. While for orthophosphate according to Table 16.1, a clear decrease, slight but measurable, occurs during this

Table 16.1 — Contamination of Rhine mountain reaches and the Upper Rhine in 1976 (I for 12–13 May, II for 28–30 September). Discharge MNQ − MQ

Sampling point		o-PO_4^{3-} (mg/l)	Total PO_4^{3-} (mg/l)	NO_3^- (mg/l)	B (mg/l)	A (mg/l)
Constance	I	0.11	0.20	3.5	0.12
	II	0.10	0.28	0.8	0.06
Stein am Rhein	I	0.17	0.25	8.0	0.07
	II	0.10	0.30	3.6	0.07
Sackingen	I	0.54	0.57	14.2	0.10
	II	0.64	0.78	10.9	0.09
Rheinfelden	I	0.56	0.68	17.2	0.10
	II	0.59	0.88	12.0	0.09
Basle	I	0.56	0.66	18.7	0.12
	II	0.50	0.90	3.0	0.10
Kembs	I	0.62	0.69	21.4	0.12
	II	0.54	0.82	3.4	0.10
Breisach	I	0.77	1.0	26.0	0.13
	II	0.70	1.5	4.4	0.10

period, in the Upper Rhine and the mountain reaches there is a drastic fall in both concentration and transport during the same period.

Trend analyses for NO_3^- ion must therefore only be performed with due regard to season and water temperature, as already mentioned in Chapter 3 and also in the text accompanying Figs 6.19, 6.20 and 6.10. Boron was also included in the Table because it also is a constituent of household detergents.

16.4 POLLUTED WATERCOURSES

Trend estimations understandably take on a major significance in connection with the study of polluted watercourses. From an overall viewpoint, the study of trends can be viewed as an extension of the calculation of mass transport rates, according to which the annual mass transport figures are compared with each other in a relatively simple fashion. Now in Chapter 15 the importance of the *actual* annual transport rates was strongly emphasised, apart from the enormous difficulty of determining this with an acceptable degree of accuracy. The best solution appeared to be to adopt representative values which are related to a particular value of discharge. It is precisely these representative values which in the light of the hydrological criteria may form the basis of trend estimation and illustration.

Since with the aid of some examples regarding the pollution of the Middle Rhine, the average (MQ) discharge was shown to be suitable as a 'representative' discharge from which transport rate estimations for several years could be made, the question of trend determination will now be considered using some examples and typical discharge values for the Moselle. In so doing, the reader should not be disturbed if, instead of the mass flux, the concentration is correlated with the discharge, as this appears to be very well suited to the problem on hand.

The discharge values for the Moselle at Cochem, according to Table 15.3, are MNQ=58.9 m^3/s, MQ=288 m^3/s and MHQ=1900 m^3/s.

The trend curve for orthophosphate based on discharges of less than the average low flow (Fig. 16.3) is very distinct evidence of a rising trend, and does not differ

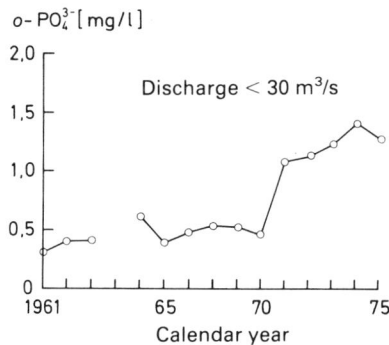

o- PO_4^{3-} [mg/l]

Discharge < 30 m^3/s

Calendar year

Fig. 16.3 — Trend for o-PO_4^{3-} concentrations in successive years in the Moselle at Koblenz, discharge < 30 m^3/s.

greatly from the curve for values standardised to 100 $m^3/s \pm 30\%$ as shown in Fig. 16.4. Both curves also reproduce the conclusions already indicated in respect of Lake Constance and the Upper Rhine.

If one now adopts the orthophosphate concentrations obtained for average discharge values, then the historical trend proves much more difficult to recognise (Fig. 16.5). Hence at least for the Moselle, the hydrological conditions obtaining for low flow regimes are much more suited to trend estimation. Average flows are suitable only for large rivers with relatively well-balanced water flows between the MNQ and MQ values, where rational trend analyses are concerned.

If one disregards the hydrological situation and instead takes the average annual mass transport of phosphate (as P) (Fig. 16.6) and correlates this with the discharge,

o- PO_4^{3-} [mg/l]

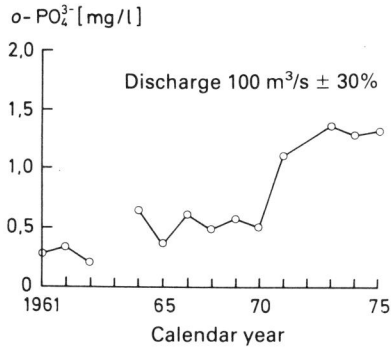

Fig. 16.4 — As for Fig. 16.3, but for value of discharge <100 m³/s.

o- PO_4^{3-} [mg/l]

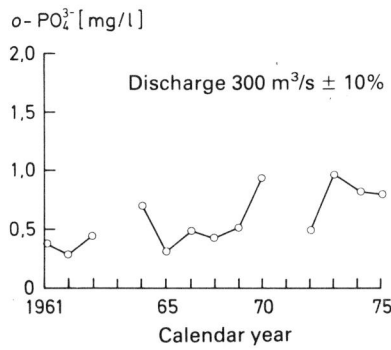

Fig. 16.5 — As for Fig. 16.3, but for values of discharge <300 m³/s (≃MQ).

P [kg/s]

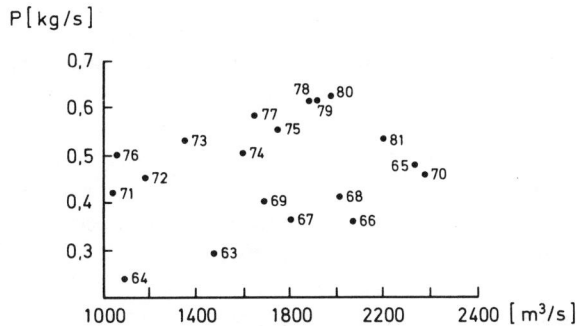

Fig. 16.6 — Rates of phosphate P transport per second in the Rhine at Koblenz, calculated from annual mean values, correlated with mean annual values of discharge.

the actual trend is 'statistically' concealed. Annual averages are therefore unsuitable, just like high reference discharge values, for trend determinations.

One should not overlook that an assessment of water quality based on low flows is not necessarily applicable to other discharge conditions. Already these few examples from the Moselle demonstrate that for average flows the orthophosphate pollution level may remain approximately constant for a period of years, quite independent of flood flows and their very difficult-to-estimate pollution levels. Where the trend curve for low flows indicates a rising or falling tendency, it still does not necessarily imply that the actual or standardised mass transport averaged over the year exhibits the same tendency.

To amplify the results already presented for concentration/discharge relationships for the Moselle in respect of orthophosphate, similar data may now be presented for the Middle Rhine. In the Rhine at Koblenz during 1959–60 the orthophosphate concentrations were chiefly in the range 0.2–0.5 mg/l (Fig. 16.7) and

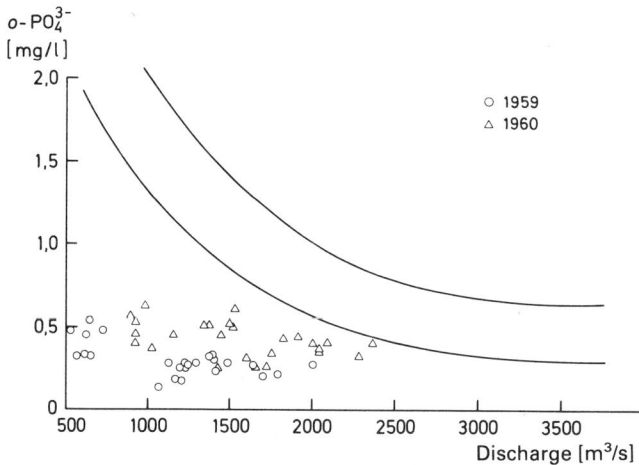

Fig. 16.7 — Concentration–discharge relationship for $o\text{-PO}_4^{3-}$ in the Rhine at Koblenz, 1959–60.

were practically independent of the discharge. In the framework of the 14-day sampling frequency of the IKSR, values for discharges in excess of 2500 m^3/s were not observed. Up to 1965/66 the orthophosphate concentrations in the low flow region showed a definite increase (Fig. 16.8). The hyperbolic decline in materials concentrations with increasing discharge approaches the lower curve. Greatly elevated values of orthophosphate concentration were observed, chiefly during the flood-prone year 1965 for discharges between 1800 and 3500 m^3/s.

A further increase up to 1978, and also for the years 1979–80 can be demonstrated (Fig. 16.9). As observed, the trend is most prominent in respect of relatively low flowrates (1000 m^3/s). If one adopts the equally favourable MQ value for the presentation of the trend data, then on the basis of the graphs the obvious question is what particular merit attaches to the values lying between the two envelope curves,

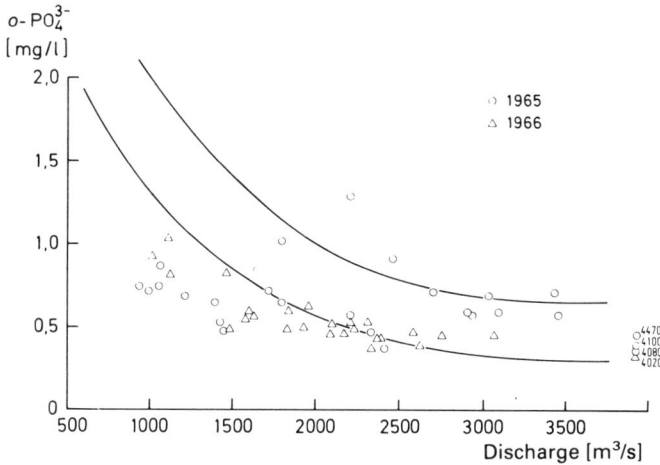

Fig. 16.8 — Concentration–discharge relationship for o-PO_4^{3-} in the Rhine at Koblenz, 1965-66.

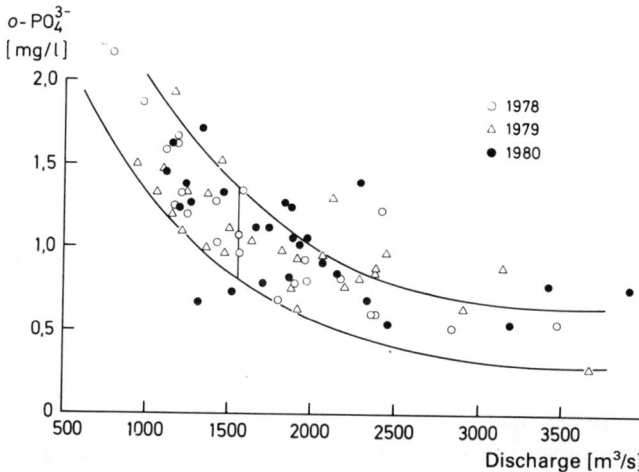

Fig. 16.9 — Concentration–discharge relationship for o-PO_4^{3-} in the Rhine at Koblenz, 1978–80.

as for the individual results, unfortunately, there is a scatter of 0.5 mg/l and over, including those for the MNQ value.

In order to be as fair as possible to the reader, the simplest of a number of alternatives was selected, although not without a certain down-to-earth approach: the envelope curves were so arranged that the majority of the values were included, but the obvious 'outliers' were excluded. Then for the MQ the 'trend value' in the shape of the average shown on the graph (Fig. 16.9) is obtained. If one then plots these values against the respective calendar year, one obtains (Fig. 6.10) the well-known indication of a rising trend in phosphate contamination, expressed as a daily mass transport rate.

The interpretation of the analytical results for MBAS (anionic detergents plus similarly reactive materials) is that much more difficult, on account of the decomposition processes in the water, which exert a varying influence on the figures owing to the changes in water temperature. At low water temperatures, the concentration/discharge relationships may be broadly represented by the curves shown in Fig. 16.11. For average flowrates in particular, daily mass transport rates were determined (Fig. 16.12), which when added together and compared with the quantities of detergents used in the catchment above the measuring station, provide positive proof of the improved rates of elimination and degradation from year to year.

For the dissolved organic carbon the ARW Report for 1976, also based on the concentration/discharge relationship, shows a definite decrease in the level of contamination between 1970 and 1974 on the one hand, and up to 1975–76 on the other (Fig. 16.13). In this case also it proved possible to base a trend on the MQ-related results, which should provide convincing proof of the improvement in the overall pollution situation for the Middle and Lower Rhine.

Further examples are to be found in Part II; thus data for oxygen are reported in Section 6.5. For zinc one may compare the information given in Fig. 7.7 and in Figs 15.2–15.4 (allowing for a conversion from mass transport rates to concentration values). In addition, ligninsulphonic acids exhibited a regionally diminishing trend (Fig. 13.6).

16.5 CONCLUDING OBSERVATIONS

The progressive change with time of specific materials concentrations and transport rates in flowing waters can be deduced from the assembled data. For this purpose a concentration or transport/discharge curve is derived for the sampling point concerned. From practical experience the degree of pollution of a free-flowing watercourse is best obtained for precisely defined discharge values in the region of the average low flow (MNQ), while for larger waterways like the Middle Rhine the corresponding values for the MQ discharge may well form the basis for calculation and presentation of results.

Although these mass transport figures are doubtless real and permit wide-ranging conclusions to be drawn in the range covered by analytical data, they should not, however, be scaled up to annual mass transport values. Numerous examples have shown that certain strongly developed pollution trends of either a positive or a negative nature observed at low flow rates may disappear with increasing values of discharge. Several causes may be advanced to explain this behaviour, although this does not seem the right place to enlarge on the matter.

The trend associated with a favourable discharge regime need not be regarded as having no connection with the 'standardised transport' (Chapter 15) but is by no means identical with it. Trends and standardised transport rates are on hydrological (and environmental–political) grounds, variously derived and evaluated. Thus, for example, in cases where the trend curve based on the MNQ indicates a falling

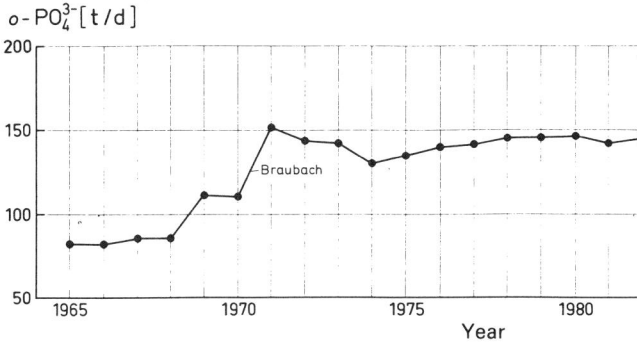

Fig. 16.10 — Progressive change in phosphate transport rates in the Rhine at Koblenz/ Braubach, standardised with respect to the mean discharge value, MQ.

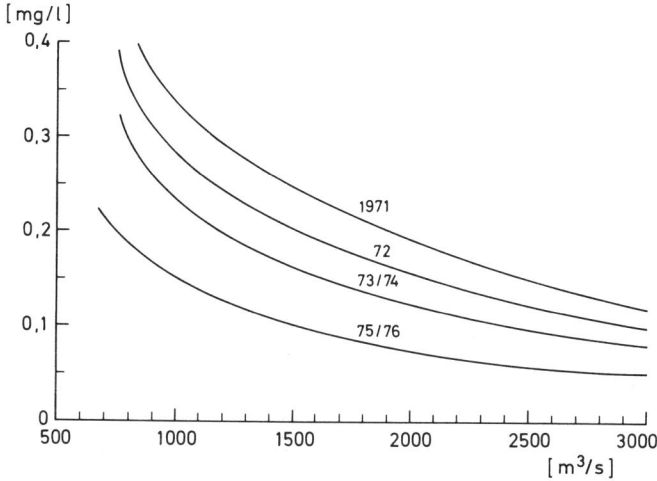

Fig. 16.11 — Concentration–discharge relationships for MBAS on the Rhine at Koblenz, standardised on the basis of water temperatures of 4–10°C.

tendency, the relevant standard annual transport rates may in fact be roughly constant.

Since, taken all round on the basis of a number of trend analyses, a positive shift in water quality has been observable since about 1975, comments on some of the less desirable happenings should perhaps not be omitted.

They concern the analysis of trace substances. As is well known, certain topical compounds and groups of compounds have little by little found their way into the province of the surface water analyst. The latest example involves the benzodioxins and benzofurans. For the exact determination of any group of trace constituents of this kind, the analyst has to acquire sufficient experience, which as a rule takes 2–3

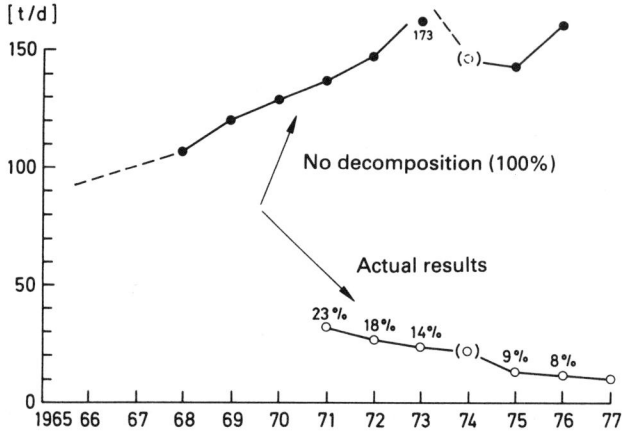

Fig. 16.12 — Trends for anionic detergent (MBAS) concentrations in the Rhine at Koblenz (lower curve), with values expected in the absence of degradation or elimination (upper curve).

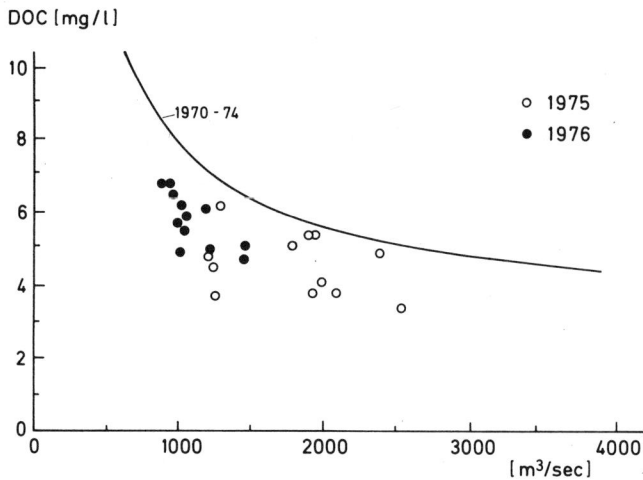

Fig. 16.13 — Concentration–discharge relationship for dissolved organic carbon in the Rhine at Koblenz (from ARN 1976).

years. In the meantime numerous analytical results get into print, which when considered in retrospect not infrequently turn out to be suspect.

These two factors have had the result that trend estimations for, e.g., heavy metals and certain organic trace contaminants must for the present be based on observations over only a few years, and in part on rather unreliable data. For the critical analyst there will thus be a special responsibility. Only he can review the development of analytical methodology and expertise and provide data of credible significance. Where the situation demands, he should not be afraid of correcting 'obsolete' values.

LITERATURE

16.2—3

Müller, G. (1977). Schadstoff-Untersuchungen an datierten Sedimentkernen aus dem Bodensee. III. Historische Entwicklung von N- und P-Verbindungen — Beziehung zur Entwicklung von Schwermetallen und polycyclischen aromatischen Kohlenwasserstoffen, Z. Naturforsch. **32** c, 920–925.

Müller, G. (1983). Zur Chronologie des Schadstoffeintrags in Gewässer, Geowissenschaften in unserer Zeit **1**, 2–11.

Bernhardt, H. (1972). Forderungen zum Schutz von Oberflächengewässern unter besonderer Berücksichtigung stehender Gewässer, ISU-Mitteilungen **1**, 37–57.

Wagner, G. (1976). Simulationsmodelle der Seeneutrophierung, dargestellt am Beispiel des Bodensee-Obersees, Arch. Hydrobiol. **78**, 1–41.

Reichenberger, E. (1976). Krautwucherungen im Rhein, Wasser, Energie, Luft **68**, 234–239.

16.4

VCI (Hrsg) (1981). Gewässergüte heute, Chemie und Umwelt; Wasser, Frankfurt a. Main; dort auch Verminderung bestimmter Einzelstoffe.

Jahresbericht der IAWR und der LWA/NRW.

Internationale Kommission zum Schutze des Rheins gegen Verunreinigung. Langfristiges Arbeitsprogramm (LZP)-Gesamtfassung, Koblenz, und weitere Berichte der IKSR, insbesondere Tätigkeitsbericht 1981.

Malle, K.-G. (1978). Wie schmutzig ist der Rhein?, Chemie in unserer Zeit **12**, 111–112.

Malle, K.-G. (1983). Der Rhein — Modell für den Gewässerschutz, Spektrum der Wissenschaft **8**, 22–32.

Groenewegen, H. J., Veltman, M. (1983). Pollution of rivers and its impact on sedimentation and marine ecology. Draft position paper, Second European environmental filmfestival, Rotterdam 22.9–2.10.83.

Hellmann, H. (1986) Schwermetalle in den Schwebstoffen des Mittelrheins 1974, 1979 und 1983/84-ein Beitrag zur Trendermittlung. Z. Wasser — Abwasser — Forsch. **19**, 85–90.

17

Sources of materials transport

17.1 INTRODUCTION

One of the most important tasks of surface water analysis concerns the quality of the water body. Besides this, the question of trend estimation, as dealt with in the preceding chapter, is also very important. Considerations of water quality in particular, as well as the degree of pollution and its long or short-term behaviour, also lead to the problem of the source of the materials concerned. An exhaustive treatment of this subject undoubtedly demands a much more intensive study than is possible within the framework of this monograph. Consequently a systematic accuont will have to give way to a few selected examples.

From a general point of view the determination of transport rates with regard to their origin requires, in addition to carefully considered sampling, a very comprehensive knowledge of the associated hydrological parameters and their effect on the results. Hence purely analytical skill should be accompanied by experience concerning the likely origin of individual transport fractions within the catchment area. One of the most important prerequisites for a realistic outcome is the preparation of a concentration/discharge relationship for the chosen measuring point, as this provides important clues regarding the source of the discharges. No less important are supportive longitudinal profile studies which should as far as possible be conducted during periods of dry-weather flow, even though investigations during wet weather and during elevated flood flows cannot be altogether dismissed.

For larger rivers it is also frequently necessary to determine the amounts originating from tributaries, a task which may be distinctly more difficult under certain circumstances than those already referred to.

If one reflects on these few comments, one might ask if a single laboratory is capable of tackling them all. If on the other hand several groups in different, physically separate, locations are all involved, then experience shows that there will be problems connected with the purely analytical aspects, quite apart from the problems of sample collection.

The difficulties roughly outlined above create a situation in which very often, even where there is a profusion of data, the actual origin of a particular mass flow remains in the dark and large-scale calculations and estimates must be employed as means of help. Under such conditions, the quality of the final outcome stands or falls with the 'correct approach' the focus of effort and the large-scale picture. Boldness of conception in the task of integrating a possible multitude of data and individual values must be combined with a conscientious and critical faculty in assessing the results. Few, but analytically sound and hydrologically well-founded, data are preferable to a large set of results riddled with possible errors or uncertainties.

A further aspect is worth remarking. For a not inconsiderable number of parameters, e.g. for heavy metals, a naturally occurring 'precontamination' of surface waters exists. This co-called background fraction must be viewed as a combination of dissolved and undissolved components and included in the further deliberations. The next section will deal more fully with the background question, with the aid of examples, than was possible in Chapters 7 and 8.

17.2 THE BACKGROUND PROBLEM

The concentration/discharge relationship at a particular point, e.g. at the frequently cited point on the left bank of the Rhine at Koblenz 590.3 km from the source, may, in the case of the natural water constituents which are not involved in chemical or biological transformation processes, be represented by a straight line parallel to the x-axis (discharge axis). The concentration of the particular substance is accordingly independent of the discharge and appreciably constant. This situation was described in Fig. 6.4 for the case of chloride in the upper Rhine at Stein. The concentration of artificially introduced substances must always decrease, normally in a hyperbolic fashion if the discharge increases and the quantity of the material introduced stays roughly constant. The combination of a constant naturally derived concentration on the one hand and a constant anthropogenic input on the other leads to analytical results for which the hyperbolic profile becomes more pronounced as the anthropogenic contribution becomes the more predominant (Fig. 17.1, top, and also Fig. 7.3).

An informative example of this behaviour can be drawn from the pattern of detergent concentration values during the 1970s (Fig. 16.11). The case is very instructive because the concentration discharge relationship undergoes an almost continuous translation from an initially hyperbolic pattern to something closer to the straight line format, during which a profile similar to that of Fig. 17.1 (bottom) emerges.

According to the statement of a moment ago, the straight line form should be evidence of a natural baseload. As, however, there are no 'natural' alkylbenzenesulphonates and the like, the baseload must be composed of interfering biogenic materials. Put another way, this means that the method of determination is lacking in specificity and the background term is consequently no longer applicable.

Also, but not solely in the case of phenols, 'background' contamination has been determined in certain river reaches which is due to nothing more than coupling-prone materials of natural origin, which have little or nothing in common with the phenols of industrial manufacture. These 'pseudo'-phenols are in part associated or incorporated with the suspended solids. If therefore one assesses the concentration/discharge relationships, these will only be meaningful if the dissolved and undis-

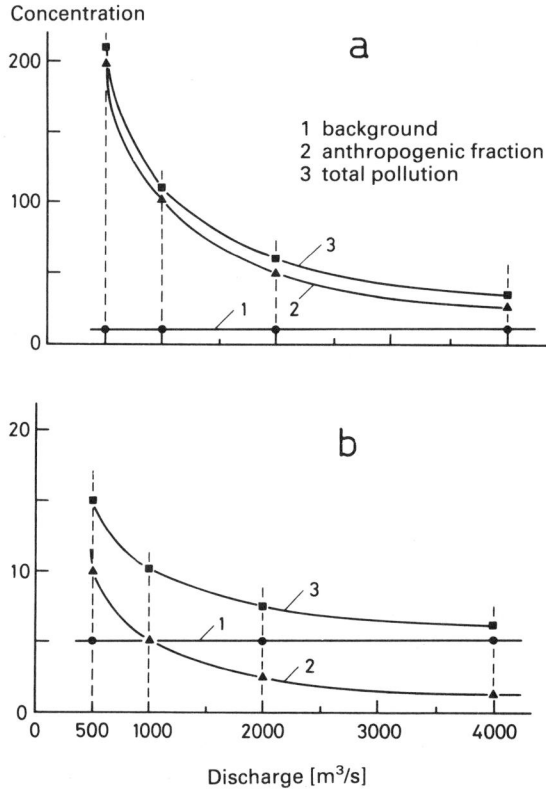

Fig. 17.1 — Concentration–discharge relationships (schematic) showing how the total pollution is formed. Anthropogenic fraction (a) relatively high, and (b) relatively low, in comparison to the background.

solved fractions are considered separately and the respective biogenic fractions distinguished.

According to these facts, when determining the source of the materials concerned one must first of all devote one's attention to distinguishing between the background and the anthropogenic components. A useful and by now 'classical' approach to this subject is illustrated by the data for undissolved chromium, i.e. chromium bound to suspended solids, in Fig. 7.6. By no means frequently does the analyst come across the practically theoretical shape of the hyperbola. From the steep decline of the specific level of contamination with the rise of discharge from 1000 to 1500 m³/s the very considerable anthropogenic component can be inferred. More precise values for the — supposedly constant — background level can in principle be deduced from the graph.

For this purpose one superimposes onto the concentration/discharge curve (Fig. 17.2) the auxiliary vertical lines at discharge values of 1000 (=Q) 2000 (=2Q) and 4000 (=4Q) m³/s. The corresponding anthropogenically defined concentrations

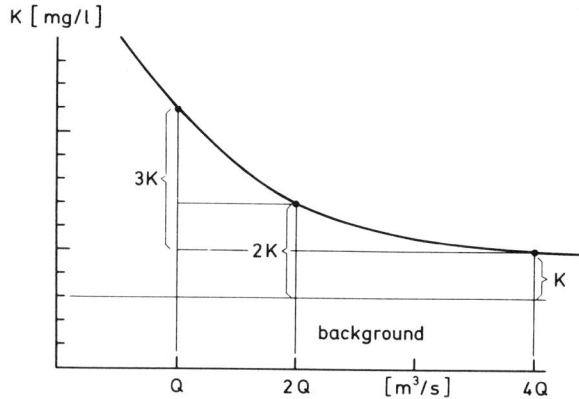

Fig. 17.2 — Graphical method of deriving the level of background pollution.

are given by the intercepts for K, 2K and 3K. The residual background value can then be read off from the graph.

Using this method, a background concentration for the sulphate anion in the Middle Rhine of 34 mg/l can be derived (Fig. 17.3) and for the Moselle at Koblenz,

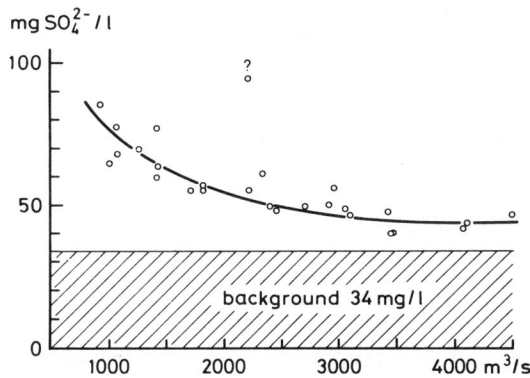

Fig. 17.3 — Derivation of the SO_4^{2-} background concentration in the Rhine at Koblenz, 1965.

37 mg/l (Fig. 17.4). Nevertheless it must not be forgotten that the size of these figures depends on the values assigned to K, which was shown in Chapters 15 and 16 to be capable of varying quite considerably for otherwise constant values of discharge. For the specific chromium contamination level, for example (Fig. 7.6), one may obtain background values which range from 30 to 125 mg/kg for different conditions.

By means of similar calculations, the background level of dissolved zinc in the Middle Rhine in 1973/74 was found to be 6 μg/l.

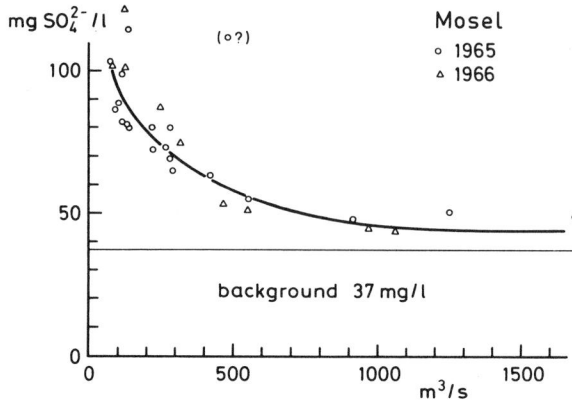

Fig. 17.4 — Derivation of the SO_4^{2-} background concentration in the Moselle at Koblenz, 1965–66.

These values are of the appropriate order of magnitude and should suffice to meet the requirements for numerous situations. In so far as the heavy metal pollution of the suspended solids is concerned, useful information may be obtained from a comparison of such values with those given by Kloke and the standard compositional data of Turekian and Wedepohl (Table 7.2). One need not stress the fact that the values for the background level are dependent on the geological composition of the subsoil in the catchment area and may therefore in some cases be relatively large.

Besides connecting the concentration with the discharge in a graphical format, the mass transport rate — as already discussed — may also be treated similarly, which may indeed be necessary, as indicated by the following chain of argument. If one considers the mass transport as a function of the discharge, then for numerous parameters the following picture will emerge in respect of flowing waters: while for low flows the natural component of the mass flow is of lesser importance relative to the anthropogenic component, it nevertheless overtakes it to a greater or lesser extent under flood flow conditions (cf. Figs 4.3 and 4.5). Once again it may be remarked at this point that the magnitude of the mass transport rate is only relevant for a particular value of the discharge. So long as the anthropogenic pollution was the predominant factor for many parameters, this point was of minor importance. However, in the meantime (by 1985) the degree of pollution of surface waters in the Federal Republic has declined to a point at which the background must once again receive serious consideration.

Lastly a problem will be mentioned which strictly belongs to the next section, but on account of a certain overlap with the background it may be preferable to include it here, viz. the question of so-called 'past pollution'. By this term one understands those deposits of heavily polluted sediments which were laid down chiefly in regulated rivers with slow-moving water. They give rise to an increase in the mass transport of many heavy metals, as well as phosphates and substances dissolved in the interstitial water, under the influence of rising flood waves and their accompanying erosive action. Past pollution in its widest sense should also in our opinion include the discarded materials 'tipped' in permeable strata, about which usually very little is known concerning either their origin or amounts. Such 'past pollution' will of course migrate via the groundwater flow into the surface water system where it will contribute to the concentration or transport/discharge relationship and impose a special slant on the interpretation of the analytical results.

17.3 ORIGIN OF MASS FLOWS IN FLOWING WATER

If one follows the train of thought represented by the diagram in Fig. 17.5, then the

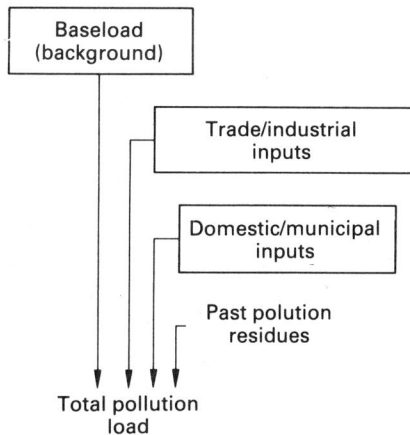

Fig. 17.5 — Composite nature of the pollutant transport in receiving streams and rivers.

total mass flow is seen to consist of four components. However, not all four components are important for every constituent. Moreover, the baseload may vary widely from time to time. A further questionable factor is the size of the input from the atmosphere, in the form of atmospheric dust and rainfall, and its influence on the analytical results.

Besides the broad subdivision of the total mass transport, the separate components may also in certain cases be further broken down. Depending on the problem in hand, it may be a question of looking for local trade waste dischargers or point-source emissions. It may also happen that the source of a particular parameter may be located in the sewage effluent discharged from a municipal treatment plant, as for phosphate, for example.

As already remarked, an exhaustive treatment of this subject is not possible here, and the experience of the author is in any case confined to the large surface water systems such as the Rhine and its tributaries.

For the most part the following reflections are based on the concentration or transport/discharge relationship, together with longitudinal profile studies. Concentration/discharge profiles for numerous substances have been graphically presented in the preceding sections and a summary of these is given in Table 17.1. Similarly summarised are mass transport/discharge profiles (Table 17.2) and concentration/ distance from source (Table 17.3).

For suspended solids in particular, some details have already been presented in Chapters 2 and 4. For these the special situation obtains that not merely the materials transport rates but also the concentrations increase exponentially with the discharge.

The size of the natural baseload is determined by the climatic conditions, the erosive action of the water and also the soil characteristics in the catchment area. From the results presented it is not difficult to see that, on the whole, the baseload contributes the major proportion and the contribution of effluent to the annual mass transport is regressing. Experience shows that extreme rainfall events on paved surfaces or hard (frozen) soils on terraces can wash away solids into the river in amounts of over 100 000 t in a few days. That this so-called 'baseload' differs from the situation obtaining in earlier years, as a consequence of urbanisation, cultivation, paving and development processes, is indisputable. However, nature is still mainly responsible for the discharge of suspended solids into the water system rather than inputs of anthropogenic origin.

For a particular catchment area the input of solids may be further examined, chiefly with the aid of longitudinal profile studies. Depending on the weather conditions, different results may be obtained. For dry weather flows the suspended solids level increases in proportion to the population of the catchment and hence to the volume of sewage effluent discharged, and also with respect to the scale of biosynthesis in the river and the erosive action of the water body itself, as Fig. 2.3 indicates for the Rhine. Rainfall events of limited extent in time and space release quantities of suspended solids which are very difficult to estimate, as a function of their intensity. If longitudinal profile studies happen to coincide with such events, then suspended solids maxima may occur at arbitrary locations (Fig. 17.6). In the Figure, it would appear that the tributaries Rems and Enz have introduced large amounts of suspended solids into the Neckar. More detailed investigation of the composition of the solids, however, shows them to be natural constituents of the soil, of which 50% is attributable to the quartz/silica fraction.

Besides the erosion-dictated suspended solids contribution, especially where surface erosion is concerned, the contribution from past waste deposits cannot be inferred without further analytical tests. From a hydrological point of view, the maxima for both fractions frequently occur together and they precede the actual flood peak. Thus, for example, the highly laden Rems contributes suspended solids to the Neckar at any given time, added to which old deposits are also mobilised within the Poppenweiler impounded reach. The most important distinction between the eroded material (bedload) and the resuspended deposits lies in the higher content of organic matter in the older deposits and in the associated higher level of contamination with trace substances such as heavy metals, and organic compounds. Analytically this gives rise to a mixture such that for relatively high suspended solids contents the specific contamination load is disproportionately increased above the level corresponding either to the baseload contamination or the soil limiting value (as in the Moselle/Saar, for example).

A certain grey area also arises as a result of the input of atmospheric dust

Table 17.1 — Review and summary list of concentration/discharge relationships

No.	Chemical parameter	River and location	Figure(s)	Remarks
1	Suspended solids	Rhine/Koblenz	2.1	—
2	Suspended solids	general	2.2	—
3	Suspended solids	general	4.3	—
4	Suspended solids	Rhine/Koblenz	4.4	—
5	Cl^-	Rhine/Stein	6.3	Dissolved
6	PO_4^{3-}	Moselle/Koblenz	6.12	Dissolved
7	PO_4^{3-}	Rhine/Koblenz	6.13	Undissolved
8	PO_4^{3-}	Rhine/Koblenz	6.14	Dissolved, total
9	PO_4^{3-}	Rhine/Koblenz	16.7–16.9	Dissolved
10	Zn	Rhine/Koblenz	7.4	Dissolved, undissolved
11	Zn	Rhine/Bimmen	7.7	Dissolved, undissolved
12	Cr	Rhine/Koblenz	7.5, 7.6	Dissolved, undissolved
13	Cd/Cu/Cr/Pb/Zn/P	Rhine/Koblenz	7.9	Undissolved
14	Cd	Rhine/Bimmen	15.3	Dissolved
15	DOC	Rhine/Koblenz	13.3	Dissolved
16	DOC	Rhine/Koblenz	16.13	Dissolved
17	MBAS	Rhine/Koblenz	16.11	Dissolved
18	$CHCl_3$	Rhine/Koblenz	11.5	Dissolved
19	AOCl/TOCl	Main	13.9	Dissolved, total

consequent on rainfall events. Over and above this there are solids introduced by stormwater overflows in the sewer systen, which may be of the separate or combined type, and on top of this there is also the effect of the poorer treatment performance at the sewage works during wet weather. Hence the generally highly polluted particulate solids derived from these sources are thrown together with the elevated bedload and resuspended sludge deposits. Only very specific investigations will serve to identify the source of these fractions in any particular case.

While both the concentration and mass transport of the NO_3^- ion in the Rhine under dry weather flow conditions increases in the downstream direction in proportion to the total population (Fig. 6.11), nevertheless the situation under different flow conditions is much more complex even at a single sampling point. Thus in the case illustrated in Fig. 6.9 the mass transport increases quite markedly, but the concentration only minimally, which would not be expected to result from a purely anthropogenic input. The explanation lies in the additional mobilisation of NO_3^- from cultivated and forested soils as a result of interflow and groundwater inputs (Figs 1.7 and 1.8). The correlation between mass transport and volumetric flowrate may be very close indeed but may alter over the years. To what extent it may be possible to make a distinction in respect of the NO_3^- ion between the effects of artificial fertilisation and the natural background contamination may perhaps be

Table 17.2 — Summary of Figures showing relationships between mass transport and discharge for specific substances

No.	Chemical parameter	Rhine sampling station	Figure	Remarks
1	Cl^-	Stein	6.5	Dissolved
2	Cl^-	Koblenz	15.1	Dissolved
3	SO_4^{2-}	Stein	6.6	Dissolved
4	NO_3^-	Koblenz	6.9	Dissolved
5	PO_4^{3-}	Koblenz	6.15	Dissolved, undissolved, total
6	P	Koblenz	16.6	Dissolved
7	Zn	Koblenz	15.2	Dissolved
8	Zn	Koblenz	15.4	Undissolved
9	KW	Koblenz	15.5	Undissolved
10	PCA	Bimmen	15.6	Dissolved, undissolved, total
11	Humic acids	—	13.7	Dissolved
12	PCB	Koblenz	10.4	Undissolved
13	HCB	Koblenz	10.5	Undissolved
14	$CHCl_3$ Tri	Koblenz	11.8	Dissolved

Table 17.3 — Summary of Figures showing concentration of substances as a function of distance from the source (longitudinal profiles)

No.	Parameter	River	Figure	Remarks
1	Suspended solids	Rhine	2.3	Mass transport
2	NO_3^-	Rhine	6.11	Dissolved
3	PO_4^{3-}	Rhine	6.16	Dissolved PO_4^{3-}, $o\text{-}PO_4^{3-}$
4	Pb	Rhine	7.11	Dissolved, total
5	Pb	Rhine	7.12	Dissolved
6	Pb	Rhine	7.13	Undissolved
7	Zn	Rhine	7.10	Dissolved, total
8	PAH	Rhine	9.10	Dissolved, total
9	PAH	Rhine	9.11	Undissolved
10	$CHCl_3/CCl_4$	Rhine	11.9	Dissolved
11	$CHCl_3/CCl_4$	Ruhr	11.10	Dissolved
12	$CHCl_3/CCl_4$	Lower Main	11.11	Dissolved

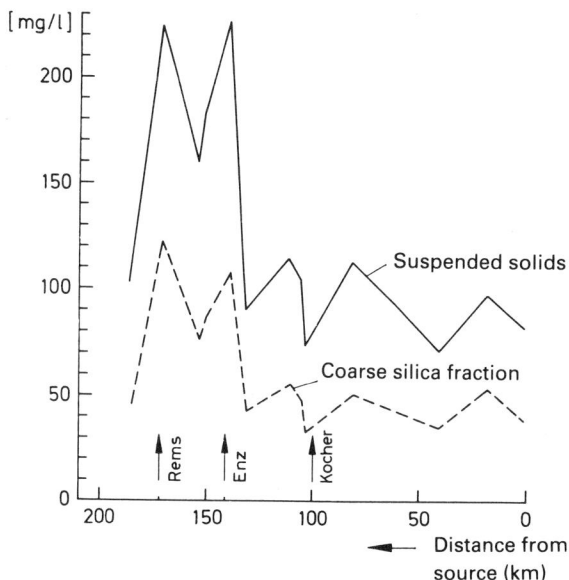

Fig. 17.6 — Suspended solids contents in the regulated portion of the Neckar, 26–29 April 1965, showing coarse silica input from tributaries.

estimated taking each case separately, but a precise demarcation between the two is hardly feasible.

In Fig. 17.7, according to the manner indicated in Section 17.1, the available analytical data are plotted in a somewhat schematic, generalised fashion. Also in Fig. 17.8, which particularly relates to medium flow conditions, a 'broad-brush' impression is provided.

Dissolved phosphates, according to reliable scientific evidence, do not originate from soils cultivated for agricultural purposes. Besides this, the actual baseload in a flowing water system is usually very small — see data for the Rhine at Constance in Fig. 17.9. This does not mean, however, that inputs from agricultural sources are invariably of negligible size. A greater degree of importance attaches to the phosphate liberated by the action of rainfall in the topsoil. In the total mass transport, contributions from human and animal excreta must not be overlooked. On top of this there is the politically sensitive topic of the contribution from household detergent products. In total, the phosphate transport at a given sampling point increases with the discharge (Fig. 6.15), while the relative importance of phosphates of faecal or domestic detergent origin declines relative to the contribution from phosphate fertilisers. The breakdown indicated in Fig. 17.8 (the position in 1974–76) is accordingly not applicable to flood flow situations.

As far as the basic hydrological facts already discussed are involved, the theme may be extended to take into consideration the so-called trace substances of the 'black' or 'grey lists'. In such cases too, the effects of past deposits, runoff from paved

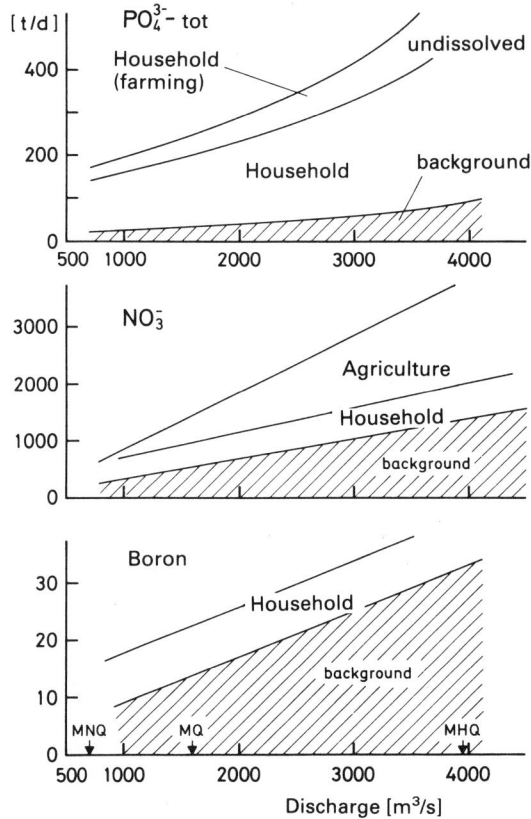

Fig. 17.7 — Origins of pollutant transport — relative magnitudes for the Rhine at Koblenz in 1974–76 as a function of discharge.

surfaces (including atmospheric washout), the discharge from stormwater overflows and sewage plant effluent outfalls must all combine to influence the result to an extent which is difficult to quantify and may call for specialised investigation.

More recent results, in the form of concentration/discharge relationships for zinc and chromium, such as those presented in Fig. 17.10 for the Rhine at Koblenz, tell the analyst that with regard to chromium the background value has practically been reached and accordingly the baseload is clearly in excess of the anthropogenic contribution. For zinc the anthropogenic inputs combined with that from past deposits are distinctly higher than the baseload of roughly 100 mg/kg; compared with the situation in the 1970s, however, the level of pollution has dropped sharply.

For many organic trace substances, in contrast to heavy metals, the possibility of background contamination can be excluded. Nevertheless the problem of identification of the source presents no small difficulties. Typical examples comprise DDT and the PCBs in addition to the HCH isomers, for besides very many non-point-source

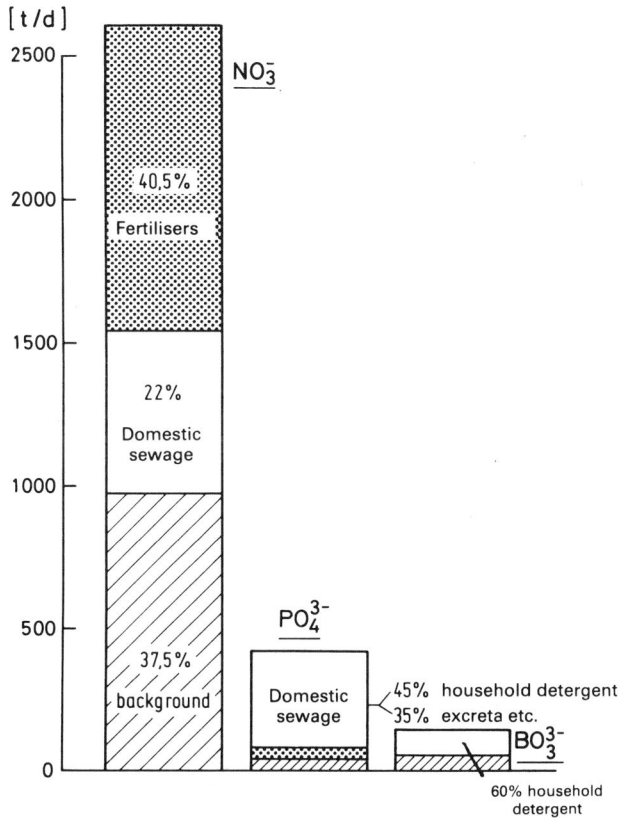

Fig. 17.8 — Make-up of pollutant transport loads in the Rhine at Koblenz, for 1974–76, under conditions of mean discharge (1560 m³/s).

inputs the effect of past deposits and atmospheric pollution must also be included. In many river reaches in the German Federal Republic, hexachlorobenzene is the compound most frequently encountered among the group of hard organochlorine compounds (1984/85 position). As HCB, on the other hand, also occurs in other surface water systems of comparable size at very much lower concentrations, the origin in the former case must be ascribed to a number of trade wastes or industrial sources.

Finally, the question of the origin of hydrocarbons in surface waters still seems largely unexplained. Besides their biosynthesis by land plants and phytoplankton, other biochemical and microbial conversion processes in suspended solids and sludges, as well as in sewage treatment plants, must be assumed to occur, which have so far not received detailed study with regard to their quantitative impact on the hydrocarbon transport in a surface water body. It may also be noted that hydrocarbons of biogenic and mineral origin are characterised by different rates of biodegradation.

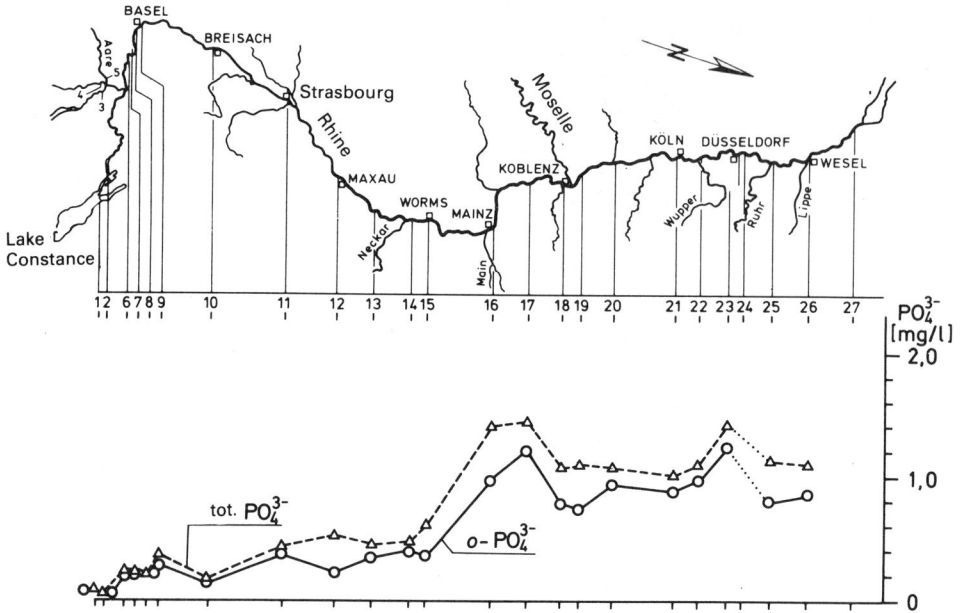

Fig. 17.9 — Longitudinal profiles for o-PO_4^{3-} and tot.-PO_4^{3-} concentrations in the Rhine from Constance to Wesel, 1–5 July 1985 (dry weather flow).

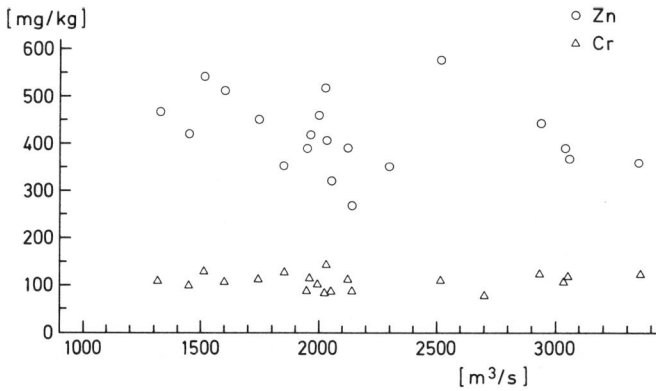

Fig. 17.10 — Levels of Zn and Cr per unit weight of suspended solids in the Rhine at Koblenz in 1983–84, as a function of discharge.

LITERATURE

Hellmann, H. (1972). Definition und Bedeutung des backgrounds für umweltschutz-bezogene gewässerkundliche Untersuchungen, Dtsch. Gewässerkd. Mitt. **16**, 170–174.

Ermittlung der Art und Herkunft der Sedimente im Rhein und in seinen Nebenflüssen im Bereich der Bundesrepublik, zu G/340.7/1100, Bundesanstalt für Gewässerkunde 1968 (Forschungsbericht).

Hellmann, H., Bruns, F.-J. (1970). Untersuchungen zur Kohlenwasserstoff-Fracht des Rheins 1968/69 und Überlegungen zu deren Herkunft — ein Beitrag zur Frage der Mineralölverschmutzung, Dtsch. Gewässerkd. Mitt. **14**, 14–18.

Hellmann, H., Griffatong, A. (1972). Herkunft der Sinkstoffablagerungen in Gewässern, Dtsch. Gewässerkd. Mitt. **16**, 14–18 und 137–141.

Hellmann, H. (1974). Zur Herkunft von Kohlenwasserstoffen in Sedimenten, Vom Wasser **43**, 179–192.

Hellmann, H. (1980). Auftreten, Herkunft und Abbau von polycyclischen aromatischen Kohlenwasserstoffen im Rhein, gwf-Wasser/Abwasser **121**, 178–184.

Hellmann, H. (1982). Polycyclische aromatische Kohlenwasserstoffe in Acker- und Waldböden und ihr Beitrag zur Gewässerbelastung, Dtsch. Gewässerkd. Mitt. **26**, 63–69.

Bilanzierung und Herkunft toxischer Schwermetalle für das Rheineinzugsgebiet (1977). Forschungsvorhaben 9/73 (W) für das Bundesministerium des Innern, BfG, Koblenz; sowie Schleichert, U. (1977/78), Berichte über Landwirtschaft **55**, 691–699.

Koppe, P., Dietz, F. (1978). Untersuchungen über die Herkunft von Schwermetallen in Gewässern, Hydrochem. Hydrogeol. Mitt. **3**, 181–202.

Auftreten und Herkunft von Zink in Gewässern (1973). Literaturbericht 1972/73, Bundesanstalt für Gewässerkunde, Koblenz.

18

Materials balances—rough estimates—model calculations

18.1 INTRODUCTION

In discussing materials balances in this chapter, a topic is addressed which has assumed an ever-increasing importance in the context of general environmental questions of recent years. The question can be concisely stated as 'Where do all the chemicals manufactured and marketed in a consumer-orientated society eventually finish up?' Many substances, after use in the intended manner, end up in surface water systems, such as detergents and phosphates. For others, their entry into the water system, either by direct inputs or via the atmosphere (hydrocarbons, PAHs), cannot be wholly prevented. Very many other compounds are formed in trace amounts in the course of industrial and trade processes. They are generally only recoverable to a certain limited extent with the aid of effluent treatment measures —among these are the heavy metals and the so-called trace organic contaminants. Other substances worthy of mention are the compounds employed for protection of people and plants, such as DDT and other herbicides and pesticides which one might not expect to find in the surface water system. Still less was it expected prior to the 1970s that the use of PCBs in largely *closed* systems would ultimately give rise to the problems that have since occurred due to the presence of these compounds in the aquatic environment.

The initial question, 'Where do the commercially produced substances go?', assumes that one has at one's disposal reliable figures regarding the quantities manufactured and 'consumed'. In evaluating the surface water analyses these figures are compared with the laboratory results. Usually one compares the annual production, or the quantity sold in any one year, with the annual mass transport rate for the surface water system. However, the analyst whose task is solely concerned with surface waters is not normally in a position to balance the equation completely. Those substances which are removed in the course of sewage or effluent treatment, or are dispersed to the atmosphere, or are deposited on the land without any appreciable contact with surface water bodies and are perhaps dissipated in the underlying soil, are naturally excluded from a mass balance calculation based on the surface water system. Moreover the demands in respect of the accuracy of the 'debit and credit' terms should not be overlooked. Neither the production and use figures, nor the materials transport rates in surface waters (Chapter 15), can be stated to an accuracy greater than ± 5–10%. However, this does not normally diminish the basic advantages of constructing such a mass balance.

It not invariably happens that the analytical data are only correct to within an order of magnitude. If, however, the calculations are based on unreliable data, there should always be due regard to the necessity of subsequent verification. This reservation must be even more heavily stressed where 'scaled up' estimates are concerned. From the scale-up to the—usually inevitable—'guesstimate' is only a small step.

The last-named variants in the field of mass balance calculations can be most appropriately termed 'model calculations'. These are exclusively performed on the basis of assumptions and provisos, the plausibility of which should on the whole stand up to critical examination but for which nevertheless no concrete analytical data are available in support.

This then is how the next few sections will be arranged, with references to typical examples of materials balances, 'guesstimates' and model calculations.

18.2 MATERIALS BALANCES

Anionic detergents

According to data from the raw materials manufacturers, which were made available to the author through the BMI/LAWA 'Detergents Committee', a total of 160 000 t of anionic detergents was consumed in 1983. If one assumes that the entire quantity is discharged to sewer, and furthermore that the anionics were composed entirely of alkane and alkylbenzenesulphonates or similarly degradable sulphates, which can be detected with the formation of the methylene-blue complex, then the balance can be constructed in the following manner.

The average discharge of the Rhine at the Dutch–German frontier amounts to 2300 m^3/s or 70.36×10^9 m^3/annum (Table 4.1). If one then assumes a similar value for the sum of all other rivers, then we obtain a total discharge of double this figure or 140×10^9 m^3/annum for the surface waters of the German Federal Republic prior to their exit from the national territory, which would mean that in the absence of any further elimination of anionics by sewage treatment or biochemical decomposition, an average (MBAS) detergent concentration of 1.14 mg/l would be expected on analysis.

On the contrary, one nowadays obtains MBAS contents, e.g. for the Lower Rhine, of <0.1 mg/l (Fig. 12.1) and the concentration of the actual anionic detergents themselves should not exceed 0.02 mg/l on average. Consequently the total flux of all the rivers crossing the national frontier is less than 2% of the relevant quantities consumed.

Phosphate

In 1974 the mass transport of total PO_4^{3-} in the Rhine at Bimmen/Lobith was around 420 t/d (Hellmann 1977), based on the mean discharge. On this basis, the annual mass transport for the Rhine and all the other rivers crossing the frontier amounts to

307 000 t PO_4^{3-}/annum or 101 000 t P/annum. The total import of phosphorus according to Table 18.1 amounted to 534 000 t/annum of which 19% could be accounted for by the surface water pollution load. Of this total, for each of the 62 million inhabitants of the Federal Republic of Germany the figure is equivalent to 4.5 g P/d. Now an adult consumes an average of 1.7 g P/d in the diet, an amount which is subsequently excreted. Part of the phosphate entering the environment is also retained in the sewage sludge, of which 2 million tonnes dry wt/annum with a P-content of 1.9% is produced, equivalent to *ca.* 38 000 t P/a. The P-balance according to Fig. 18.1 thus leads to a residue, from the standpoint of the surface water analysts, of 395 000 t P/a of phosphate (1973–76), the fate of which is unknown. On the other hand it is known that considerable quantities are stored within cultivated soils. In flood-prone years the process of surface erosion can lead to an increase of 10–20% in the average mass transport of P for that year.

Low-boiling chlorinated hydrocarbons

The annual amounts used of the quantitatively most important chlorinated solvents (Table 18.2) are variously reported in different places. One may safely assume, however, a total consumption of the order of 300 000 tonnes/annum and furthermore that none is recycled to the manufacturer.

According to the analytical findings (Hellmann, 1984) the five chlorinated hydrocarbons are already present in minute amounts in spring waters. In flowing waters a major proportion is lost by evaporation to the atmosphere, so that the amounts detectable by analysis are no longer comparable with the inputs. Despite the not inconsiderable fluctuations in concentration, especially for CCl_4, the following fluxes were calculated for the Rhine in 1983 at Wesel, i.e. practically on the border with the Netherlands (MQ = 2300 m^3/s):

Dichloromethane[a]	2 μg/l	145 t/annum
Trichloromethane	3 μg/l	217 t/annum
Tetrachloromethane	0.4 μg/l	29 t/annum
Trichloroethylene	0.5 μg/l	36 t/annum
Perchloroethylene	0.8 μg/l	58 t/annum
Total		485 t/annum

[a]Figures for this compound should be treated with reserve; they are probably too high.

The total mass of chlorinated hydrocarbons transported annually from West German territory into the waters of the North Sea in 1983 was thus less than 1000 t and hence represented a fraction of less than 1% of the quantities actually used. More than 99% of the total consumption must be supposed to be present in parts of the environment outside the surface water network.

Hydrocarbons

The danger that surface and groundwater systems would become contaminated by petroleum hydrocarbons first presented itself in the 1960s and 1970s. That this fear was not without justification could be seen from the rapid increase in the imports of

Table 18.1 — Distribution of phosphorus imports between various usages in the Federal Republic of Germany for 1973–75 (from Bernhardt 1978)

Usage	Amount (t/annum)	Percentage
Fertilisers	397 000	74%
Feedstuffs	31 000	6%
Technical products	106 000	20%
(including detergents and cleaning agents)	(69 000)	(13%)
Total imports	534 000	100%

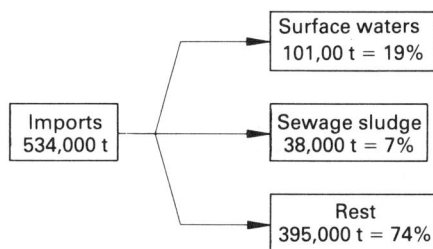

Fig. 18.1 — Phosphate (P) balance for the German Federal Republic, 1973–75.

Table 18.2 — Consumption of some low-boiling chlorinated hydrocarbons in the Federal Republic of Germany (data for 1979–84)[a]

Solvent	Consumption (t/annum)	Literature ref.
Dichloromethane	66 000	VCI 1983
Trichloromethane	50 000	VCI 1984
Tetrachloromethane	150 000	VCI 1984
Trichloroethylene	130 000	UBA 1979
Perchloroethylene	300 000	Triebig 1981

[a]Other sources indicate a total annual consumption of 300 000 t.

crude oil and hence in domestic oil consumption (Table 18.3). A classification by product types can be seen in Table 18.4.

In contrast to the situation for detergents, a use-related input to surface waters is not implied, as by far the largest proportion is accounted for by energy production, and only a very small part is devoted to lubricants. Nevertheless it would be unrealistic to assume that the input of petroleum oil to surface waters would be zero. Much rather the following potential sources of contamination must be borne in mind: rainfall and atmospheric dust (Fig. 18.2), inputs from paved surfaces via sewers and stormwater overflows, and unavoidable losses of lubricating oils, such as during boat navigation, as well as occasional oil spills on and beside surface waters. For the levels of contamination of surface waters from groundwater and from oil spills or accidental discharges on land, there are at present no firm data. They should, however, overall be very low. Based on estimates performed by different expert panels and specialists, including the Environment Ministry, scaled-up hydrocarbon emissions in the context of energy production (traffic by self-propelled vehicles, heat generation and electricity generation) amount to 1.6×10^6 t/annum, giving a theoretical level of contamination in rainwater (200×10^9 m^3/annum) of 8 mg/l. This would of course represent a considerable input to the surface water system.

Contrary to this estimate, the concentration in rainfall for 1973–75 was below 0.1–0.3 mg/l (Hellmann *et al.,* 1976, Fig. 18.3). Even this relatively low value must include a component of biogenic origin.

Apart from these considerations, the Rhine alone in 1983–84 transported an amount of 11 000 t/annum of hydrocarbons, of which 4000 t could be ascribed to the undissolved fraction. The total pollution of inshore waters originating from rivers (max. 22 000 t/annum) has thus not reached 0.2% of the quantities of petroleum hydrocarbons consumed. That possibly 50% of the values quoted may in fact be attributable to hydrocarbons of biogenic origin does not alter the order of magnitude of the figures concerned.

It is quite a different story in eutrophic freshwater lakes! In such situations the biogenic hydrocarbon formation may outweigh the input of petroleum hydrocarbons several-fold, as will be seen in the next section.

Heavy metals
Where heavy metals are concerned we are dealing with a group of substances which—in contrast to the detergents—do not enter surface waters or sewage in the course of 'correct usage'. Rather we are dealing with a residue which is a consequence of the presently imperfect state of effluent treatment for trade and industrial discharges. An alternative pathway is provided by land disposal.

According to the diagram of Fig. 18.4 the annual export of zinc across the Dutch border in the waters of the Rhine is composed of five fractions, viz. the inevitable background, the anthropogenically derived zinc content of rainwater (estimate), the unavoidable corrosion of galvanised water pipes, and to a lesser extent the zinc content of human and animal excreta and the contribution from trade and industrial effluents. The total mass transport calculated by adding these components, equal to 1935 tonnes for average discharge in 1979, is considerably below the observed value of 7700 t. Over 5000 t must therefore be derived from industrial or trade effluents, whether in the form of direct or indirect discharges. In so far as the chemical industry

Table 18.3 — Crude oil imports into the Federal Republic of Germany, 1950–84[a]

Calendar year	1950	1960	1970	1980	1984
Imports (10^6 t)	2.2	23.3	98.8	97.9	66.9

[a]From *Fakten u. Faktoren* (1985), Esso AG, Hamburg. Added to this, petroleum product imports in 1984 were at a level of 43.8×10^6 t/annum.

Table 18.4 — Classification of crude oil consumption in the Federal Republic of Germany, 1984[a]

Usage and consumption	Amount (10^6 t/annum)	Percentage
Petrochemical products	12.3	11.6
Petrol fuels	23.6	22.3
Diesel fuels	14.0	13.2
Light heating oil	33.9	32.0
Heavy heating oil	10.6	10.0
Lubricants	1.0	0.94
Bitumen	2.9	2.74
Other products	7.5	7.1
Total	105.8	100

[a]Source: as Table 18.3.

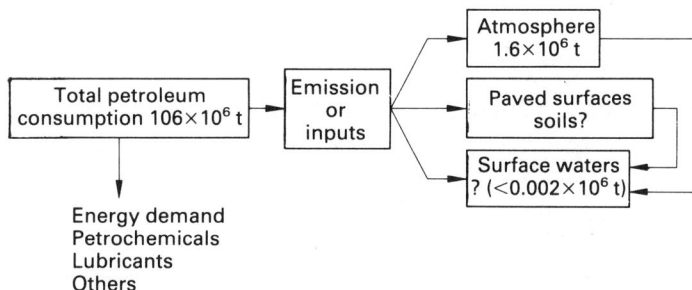

Fig. 18.2 — Consumption and fate of petroleum products in the German Federal Republic, 1984.

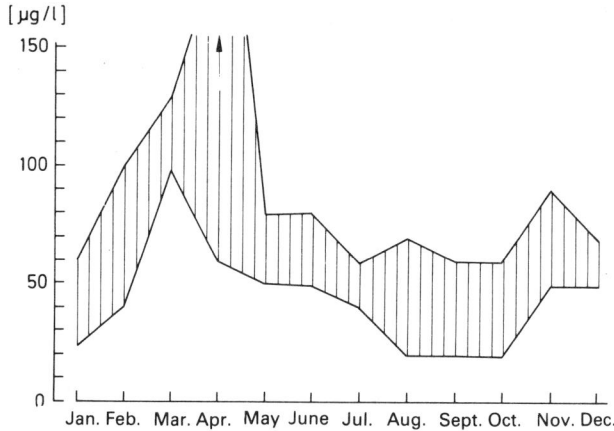

Fig. 18.3 — Hydrocarbon concentrations in rainwater at Koblenz; monthly minima and maxima, 1975.

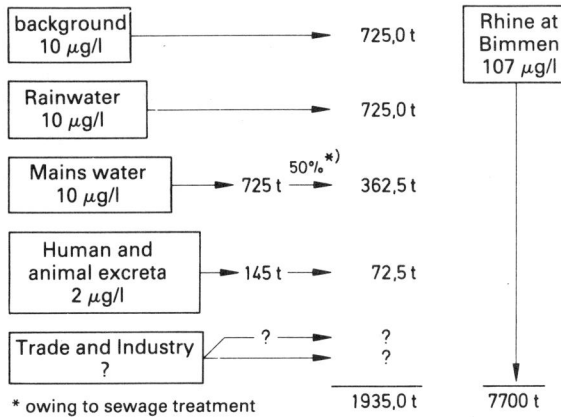

Fig. 18.4 — Tentative Zn-balance for the Rhine basin 1979 – figures in t/annum.

is responsible for direct discharges, one obtains from Table 18.5 a value of 1100 t/annum for 1985, which of course refers to the whole of West Germany. This figure is quite close to the values obtained in respect of the background component and plumbing systems. The zinc flux of the Middle Rhine from our own observations for 1983–84 can be put at 2000 t/annum; for the Lower Rhine we estimate 4500 t/annum. According to the current situation, therefore, the contributions from the background, from atmospheric inputs and from trade and industrial sources are all roughly of equal orders of magnitude.

For chromium, direct inputs in 1983 seem to have been reduced to the level of background contamination in the Rhine catchment (Table 18.5). On top of this a contribution from past pollution was observable from time to time.

Table 18.5 — Heavy metal contamination of West German waters by direct discharges from the chemical industry, tonnes/annum[a]

	1980	1983	1986 (proposed)
Lead	60	40	30
Zinc	1100	700	500
Copper	120	100	80
Nickel	60	100	80
Chromium	—	200	150

[a]From VCI Working Party on Environment, 26 April 1985.

Polycyclic aromatic hydrocarbons

Any materials balance for these compounds is practically impossible, because the 'production figures' are unknown. It is quite definite, however, that the contribution from *unused* mineral oil products is negligible (Fig. 18.5), and similarly for the

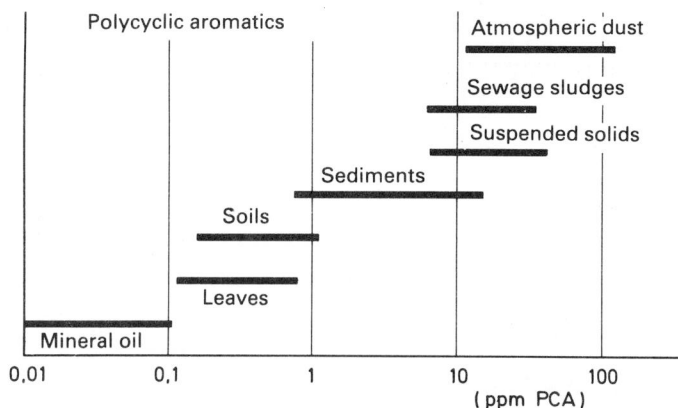

Fig. 18.5 — Polycyclic aromatic hydrocarbons (PAH) in mineral-oil products and in surface water-related solids.

scientifically rather controversial PAHs of biogenic origin. By far the largest amounts of PAHs arise from the combustion of fossil fuels as sources of energy (petrol, oil, wood, coal), as a result of which the high specific contamination of atmospheric dust and also (due to the effect of such dust) of many sewage sludges becomes understandable. The PAHs detected in surface waters originate as far as the dissolved fraction is concerned from the prior contamination of groundwater (1.5 t/annum in the Rhine at Bimmen/Lobith). The many times greater quantities of undissolved PAHs (70–100 t/annum in the Lower Rhine) should arise as a result of inputs from the atmosphere and from paved surfaces and also other sources such as ship navigation. On the other hand the amounts of PAHs transported by the waterways of the German Federal Republic amount to less than 1% of the entire quantity emitted to atmosphere by combustion, even if that amount is still very much a tentative estimate.

18.3 ROUGH ESTIMATES

In this section we are concerned with 'guesstimates' which are based partly on a few measurements and partly on assumptions. The object lies in data for certain comparisons which will not be described in detail here, and also in deriving specific accounts and total estimates as a basis for discussion, and thus occasionally to advance the process of searching for the truth.

Polychlorinated biphenyls in the Rhine

The specific PCB contamination of the suspended solids in the Lower Rhine may, in line with Table 10.1, be taken as 500 mg/t, and the mean annual transport of suspended solids as 3.5 million tonnes (Table 2.2). From this the mean annual transport of undissolved PCBs amounts to 1.75 t. The contribution of the dissolved PCBs may be assumed to be similar, giving a total annual mass transport figure for the Rhine of 3.5 t which, when the other rivers are taken into account (Elbe, Weser, Ems, Danube), implies a total figure for West Germany of 7 t/annum. This volume contrasts with a total production figure quoted by the 'German Commission' (1980) of 7500 t/annum, of which 3124 t was disposed of internally as follows:

Hydraulic oil for mining	*ca.* 1160 t ≃ 37%
Transformers	*ca.* 1100 t ≃ 35%
Small capacitors	*ca.* 530 t ≃ 17%
Large capacitors	*ca.* 350 t ≃ 11%

The main producer of PCBs, Bayer AG, according to its own published work (*Environmental Protection in Bayer*) ceased manufacture during 1983.

Hexachlorobenzene in the Rhine

The HCB content of suspended solids in the Middle Rhine at Koblenz fluctuates widely (Table 10.1). Starting from a figure of 100 mg/t, as well as other similar figures as in the case of the PCB summation, one obtains for undissolved HCB at the national frontier (Bimmen/Lobith) 0.35 t/annum with a roughly equal value for the dissolved fraction, giving 0.70 t/annum, which is doubled for the entire output to coastal waters, giving a maximum of 1.4 t/annum.

Hexachlorobenzene production and use in 1980 amounted to 4000 t (VCI 1981). According to the VCI document, in 1980 around 0.6 t of HCB were discharged from manufacturing premises to surface waters. In the same place a value of 1.2 t/annum is quoted for the annual HCB transport rate in the Rhine.

Lindane in the Rhine

Lindane, for which the concentration in Rhine suspended solids should not exceed 10 mg/t, is accordingly transported into the North Sea in Rhine water in amounts of 35 kg/annum in undissolved form and a maximum of 35 kg in dissolved form. By doubling, a value of 140 kg/annum for the total emissions from the German Federal Republic in surface waters is obtained. Production and use of lindane at the present time is far in excess of 100 t/annum. According to the DFG (1982) about 250 t are produced annually in the German Federal Republic. Apart from this the production total for the group of insecticides, according to information supplied by the Industrial Association for Plant Protection, amounted to 45,000 t in 1984 (*Nachr. Chem. Techn.*).

Contribution of atmospheric dust to the suspended solids pollution of natural waters

In the preceding chapters the question was directly or indirectly posed several times regarding the extent of the possible contribution of atmospheric inputs to the contamination of surface waters. many individual investigations have demonstrated that atmospheric dust exhibits relatively high concentrations of certain substances. These naturally enter to some degree into the total suspended solids pollution so that the quantity of this input should be capable of estimation from a materials balance. on the basis of the figures reproduced in Table 18.6, over 1 million tonnes of

Table 18.6 — Dust incidence on paved surfaces[a] and surface water bodies[b]

Paved surfaces	Area (km^2)	Dust[c]
Roads and railways	11 150	
Buildings and courtyards	10 480	500 kg/ha/annum
Industrial areas		
Airports and parade grounds	2170	$\approx$50 t/km^2/annum
Total	23 800	1.2×10^6 t/annum
Surface water bodies	4500	0.22×10^6 t/annum

[a]From Materialen/BMI (position in 1970).
[b]Author's estimate.
[c]From Reports/UBA, 500–1000 kg/ha.

atmospheric dust enter West German surface waters annually (over the cultivated land area — 140 000 km^2 — a further 7 million tonnes are deposited). That is more than 10% of the suspended solids load of the flowing waters. As the contamination of atmospheric dust with respect to certain trace substances is about one power of ten greater than that of the suspended solids occurring *in situ*, the determination of the origin of the respective substances must take into account the contribution from the atmosphere both as rainfall and as dry deposition.

To the extent that the emission situation for West German water bodies has already been improved with regard to trade effluent discharges, the atmospheric input begins to assume an increasing importance. In this connection, one quotation may be included: 'Since 1971 when the ban on DDT became effective in Germany,

DDT has been detectable solely in the dust carried by the atmosphere' (Borneff, 1973).

Added to the direct inputs from the atmosphere, there is also the indirect input resulting from the erosion of agricultural land which finds its way eventually into the surface water system. The proportion of plant derived emissions, which are noticeable in the case of the metals zinc and lead, may also not be neglected in areas of low atmospheric pollution, although from a global standpoint they may have little or no significance.

Hydrocarbon balance for Lake Constance (1975)
The situation represented diagrammatically in Fig. 18.6 may be amplified with the

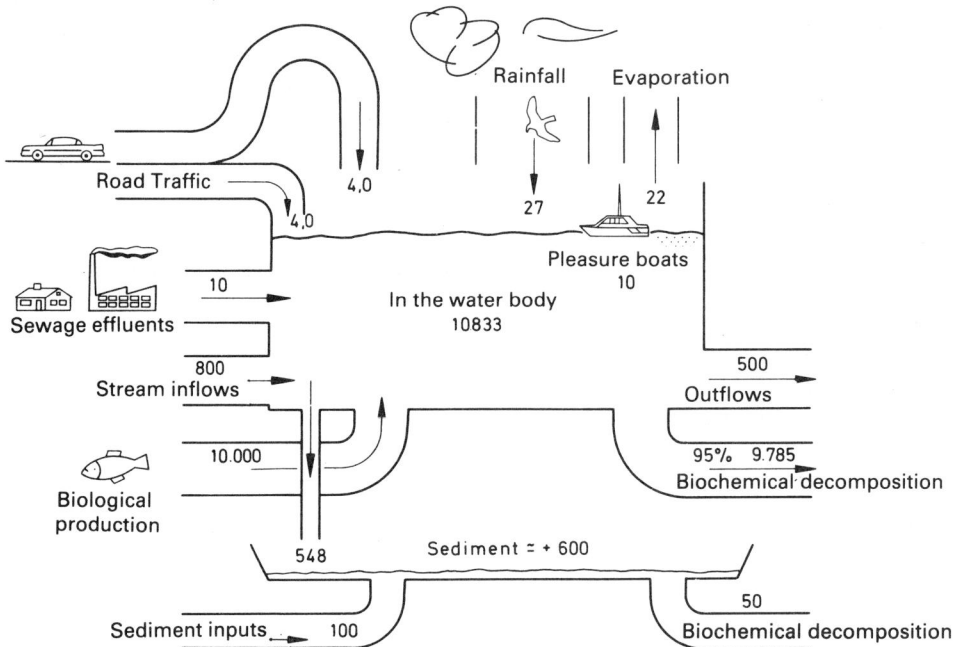

Fig. 18.6 — Trial balance for hydrocarbons in Lake Constance, 1975 (data in tonnes).

aid of the following detailed calculations (data expressed in tonnes):

1. *Evaporation*
   800 mm/annum, total surface 540 km^2
   Hydrocarbon content 50 mg/m^3
   Net loss = 21.6 t, say: −   22

2. *Wet deposition (rainfall)*
   1000 mm/annum, total surface 540 km^2 Hydrocarbon content 50 mg/m^3
   Net gain = 27 t: +   27

3. *Traffic*
   Highway length 150 km, width 10 m
   $\therefore$ Area = 15×10^5 m^2 of which 10% is assumed covered by
   an oil film 0.5 μm thick
   $\therefore$ Oil volume = 0.08 m^3. Annual washdown assumed
   equal to 50 times input (rough estimate)
      Net gain = 4 m^3:   +   4

4. *Domestic and industrial effluent*
   Total pop. 500 000 PE, *per capita*
   vol. = 200 l/d. Total vol. = 365 m^3/annum
   Hydrocarbon content 0.2 mg/m^3
      $\therefore$ Net gain = 7.3 t, say: +   10

5. *Rivers and streams*
   Inflow 300 m^3/s $\simeq 10^{10}$ m^3/annum
   Hydrocarbon content 80 mg/m^3
      $\therefore$ Net gain = 800 t: +   800

6. *Pleasure boats*
      ca. 10 t: +   10

7. *Biosynthesis*
   2×10^6 t biomass containing 0.5% hydrocarbons
      $\therefore$ Gain = 10 000 t: 10 000

               Total for 1–7   +10 829
                                                    – – – – –

8. *Outflow*
   300 m^3/s $\simeq 10^{10}$ m^3/annum
   Hydrocarbon content 50 mg/m^3
      $\therefore$ Net loss = 500 t: –   500

9. *Biochemical degradation*
   95% of 10 300 t
      $\therefore$ Net loss = 9785 t   – 9 785

               Total for 8–9   –10 285
                                                    – – – – –

               Overall net gain   544

10. *Input to sediment*   544

11. *Additional sedimentary input*
    including hydrocarbons formed by conversion of
    organic matter in sediments   +   100

12. *Biochemical degradation in sediment*

                                                                                   −   50
    Hence the annual accumulation of hydrocarbons   − − − − −
    in sediments of Lake Constance =   *ca.* 600 t
                                                                                   − − − − −

18.4   MODEL CALCULATIONS

The final stage in the interpretation of the results of surface water analyses may consist of model calculations. These may also be of benefit at or even before commencement of the actual analytical work.

In the first place they serve to place the findings with respect to the surface water system into a larger perspective, as part of the attempt to relate them to the living and consumption habits of the population, with the nature and extent of manufacturing and trade processes, as well as the behaviour of the non-aquatic environment. The few examples already mentioned in Section 18.3 provide some pointers in this direction.

In the second place one endeavours, from a knowledge of the production statistics for certain materials, or possibly from the general scale of particular emissions, to predict the probable level of contamination of natural waters. As during the years from 1950 onwards, against a background of steeply rising product development and consumption on a global scale, the study of *applied* environmental analysis has not kept pace with the safety of the general population, we are in the position of attempting to name, in retrospect, both the identity and the concentrations of all the important anthropogenic constituents of natural waters. A classic example of this situation is provided by the polychlorinated biphenyls.

The subject of national and supra-national environmental politics, especially in the EEC, is concerned with identifying problems of this nature in their initial stages and where necessary instituting countermeasures. The criteria according to which potentially toxic materials were assigned either to Annex 1 (Black list) or Annex 2 (Grey list), implying the need for particular caution, were persistence, bioaccumulation and toxicity (EEC Directive 76/464). Of the 1500 substances produced and employed for technical purposes in the European territories which were considered to warrant further expert opinion, the annual quantity produced or the annual usage were also taken into consideration. For 500 substances the specified value of 500 t/ annum was exceeded. On subsequent reflection (Malle 1984) 129 substances remain to be considered.

In these current efforts, the task of model calculation, as understood here, has a part to play. The diagram shown in Fig. 18.7 is based on an annual production figure of 1000 tonnes. Normally losses are to a certain extent unavoidable, and in the diagram they are taken as 5% or 50 t/annum. The subsequent sewage treatment process would be assumed to achieve 90% elimination, by either biological or

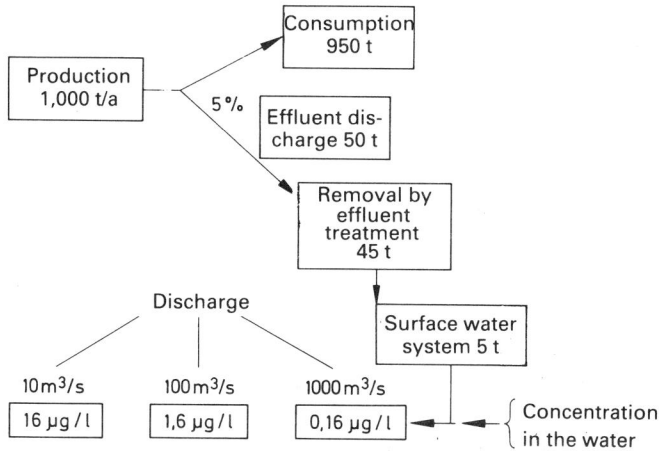

Fig. 18.7 — Model calculation for potential contamination of surface waters for stated production figures.

physico-chemical methods, with the result that a residue of 5 t inevitably finishes up in the surface water system. The average concentration in the water is then entirely a function of the discharge. A relatively small river like the Wupper (13.7 m^3/s for the annual series 1932–65) would thus exhibit a concentration of 10 μg/1, while a larger river with a discharge of 1000 m^3/s (e.g. the Rhine at Rheinfelden) would exhibit a figure of only 0.16 μg/l. The detection and estimation of materials at these levels must be achieved using the present state of analytical techniques, and moreover the behaviour of the residues with respect to their accumulation and persistence in the aquatic environment must also be pursued.

Not less current, but less securely based, are model calculations relating to atmospheric emissions, for which the hydrocarbons may serve as an example. According to the diagram in Fig. 18.8 it is assumed that 50% (0.5×10^6 t) of the total emission remains over the German Federal Republic and is present there in the form of aerosols, dust and wet deposition. (There is still unfortunately very little known about the extent to which they and other emissions are decomposed or chemically reacted in the atmosphere; consequently this very important aspect is left on one side.) As about 50% of the rainfall over the Federal German Republic escapes as direct surface runoff, it carries with it an estimated quantity of 25 000 t/annum, which would imply (for a total discharge for all the river systems to coastal waters of 100×10^9 m^3 — table 4.1) a final concentration of arounf 0.24 mg/l. The major remaining fraction of 0.475×10^6 t would then be distributed within the terrestrial environment. Regarding the percentage which, following desorption, eventually appears in the groundwater, a figure of 2% is assumed, although even the order of magnitude is open to doubt, and even more so the question of how soon and how much would make its appearance in the surface water system. That atmospheric pollutants chiefly enter the surface water system via groundwater no longer needs further proof. However, the extremely variable soil conditions below ground

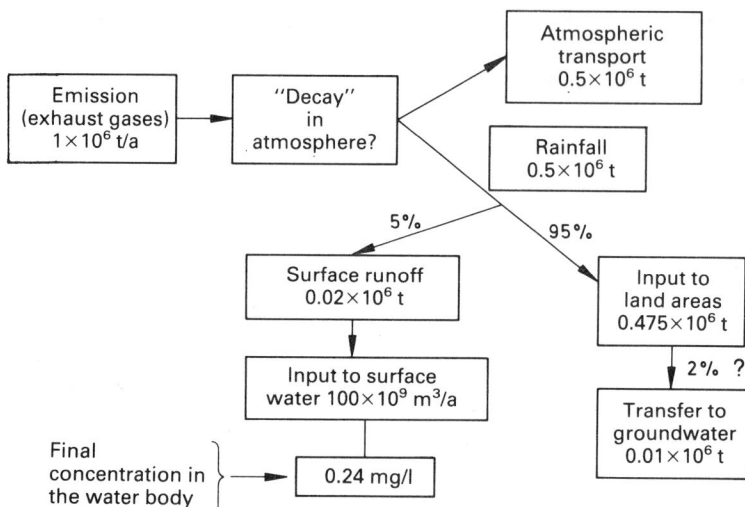

Fig. 18.8 — Model calculation for potential contamination of surface waters due to known exhaust gas emissions.

generally render the question of model calculation impracticable (see Leitfaden 1983).

Also of considerable difficulty is the question of the extent of possible contamination of surface waters by pesticides. Although their application is confined to the land area, nevertheless their input to surface waters, either in the form of atmospheric dust or via groundwater or direct surface runoff, cannot be ignored, as experience shows. Every year the Earth receives rainfall totalling 1.2×10^{14} m^3 (Hofius 1981). Based on the assumptions that half of this becomes runoff, and that 0.1% of the amount of DDT produced annually (0.1% of 150 000 t, or 150 t) is carried by the rainwater into either the groundwater or surface water systems, one arrives at a concentration of 2.5 ng/l. The concentrations currently observable in the freshwater systems of the German Federal Republic are of the order of 0.2 ng/l, and for the North Atlantic around 0.01–0.05 ng/l (1981).

Model calculations of this kind may thus furnish the analyst with an indication of the concentration range within which a particular substance might be found to occur. It may happen that, in the light of the analytical results, some of the calculations have to be corrected. Where the relevant figures are entered at the appropriate places in the calculations, then we have once again strayed into the domain of materials balances.

LITERATURE

18.2

Henkel KGaA, ZR-AS Volkswirtschaftliche Abteilung (1984). Weltproduktion und -verbrauch an Seifen, Wasch- und Reinigungsmittel 1980, 1981 und 1982, Düsseldorf.

Statistisches Bundesamt Wiesbaden (1976) Produzierendes Gewerbe, Fachserie 4, Reihe 3. Produktionsstatistik von Waschmitteln und Maschinengeschirrspülmitteln.

Hellmann, H. (1977). Zur Phosphatbelastung des Rheins in gfewässerkundlicher Darstellung. gwf-Wasser/Abwasser **118**, 259–263.

Bernhardt, H. (1978). Phosphorverbindungen und Gewässer, Wasser Berlin '77, Kongressvorträge, Colloquium Verlag Berlin, S. 181–185.

Fachgruppe Wasserchemie (1980). Wege und Verbleib des Phosphors in der Bundesrepublik Deutschland, Studie im Auftrag des Bundesministeriums des Innern.

Thormann, A. (1977). Klärschlamm-Menge und Beseitigung in der Bundesrepublik, Korrespondenz Abwasser **24**, 212–214.

Hellmann, H. (1984). Leichtflüchtige Chlorkohlenwasserstoffe in den Gewässern der Bundesrepublik Deutschland – Auftreten und Bilanz, Haustechnik — Bauphysik — Umwelttechnik — Gesundheits-Ingenieur **105**, 269–278.

Verband der Chemischen Industrie e. V., Arbeitsaussachuß Abwasserfragen. Dichlomethan-Statement vom 6.10.1983.

Verband der Chemischen Industrie e. V., Arbeitsausschuß Abwasserfragen. Statements zu Chloroform und Tetrachlorkohlenstoff vom 10.1.1984.

Umweltbundesamt (1979(. Umweltforschungsplan des Bundesministeriums des Innern. Gutachten zur Umweltverträglichkeit von ausgewählten Produckten der Chemischen Industrie, Berlin.

Triebig, G. (1981) Neue Aspekte zur Beurteilung einer Einwirkung von halogenkophlenwasserstoff-Lösemitteln aM Arbeitsplatz. Eine Literaturübersicht, Zentralbl. Bakteriol I. Abtl. Orig. B **173**, 29–44.

Haagen, H. (1984) Umwelt und Arbeitsstoffe, Vortrag Generaldebatte 1984 des Werkstoffausschusses des Hauptverbandes des deutschen Maler- und Lackiererhandwerks. Lack im Gespräch, Frankfurt am Main; dort weitere Zitate.

Immissionssituation durch den Kraftverkehr in der Bundesrepublik Deutschland (1974). Schriftenreihe des Vereins für Wasser-, Boden- und Lufthygiene 42, Gustav-Fischer-Verlag, Stuttgart.

Hellmann, H., Holeczek, M., Zehle, H. (1976). Ortganische Stoffe im Regenwasser, Vom Wasser **47**, 58–79.

Bundesanstalt für Gewässerkunde (1973). Auftreten und Herkunft von Zink Gewässern. Literaturbericht 1972/73, Koblenz.

18.3

Deutsche Konnission zur Reinhaltung des Rheins (1980). Arbeitsausschuß "Abwasserfragen", Düsseldorf.

Umweltschutz bei Bayer, 3 (1983), August-Heft.

Verband der Chemischen Industrie e. V., Arbeitsausschuß Abwasserfragen, HCB-Statement vom 30.7.1981.

Deutsche Forschungsgemeinschaft (1982). Hexachlorcyclohexan-Kontamination. Ursachen, Situation und Bewertung, Harald Boldt Verlag, S. 62.

Pflanzenschutz-Weltmarkt (1985). Deutsche halten 25 Prozent, Nachr. Chem. Techn. LaB. **33**, 668.

Bundesministerium des Innern (1971). materialien zum Unweltprogramm der Bundesregierung, zu Drucksache VI/2710, S.9.

Röber, H. M., Höllwarth M. (1984). Schwermetallbelastung durch Regenüberläufe, Haustechnik-Bauphysik-Umwelttechnik-Gesundheits-Ing. **105,** 45–49.

Umweltbundesamt (1985). Deposition von Luftverunreinigingen in der BRD — Stand Mitte 1984, Berichte 4/85, Erich Schmidt-Verlag, Berlin, S. 4–13.

Borneff, J. (1973) Die Verunreinigung von Grund- und Obereflächenwasser durch organische Substanzen des Luftaerosols, Schriftenreihe der Vereinigung Deutscher Gewässerschutz e. V. VDG Nr. 32/1973.

Institut für Wasser-, Boden- und Lufthygiene des Bundesgesundheitsamtes, zahlreiche Schriften der port herausgegebenen Schriftenreihe und andere Publikationen.

Thomas, W., Rieβ, W., Herrmenn, R. 1983) Processes and Rates of deposition of Air Pollutants in Different Ecosystems, in Effects of Accumulation of Air Pollutants in Forest Ecosystems, D. Reidel Publishing Company, S. 65–82.

18.4

Holman, W. F. (1981). Estimating the Environmental Concentrations of Consumer Product Components, Special Technical Publication 737, American Society for Testing and Materials. 1916 Race Street, Philadelphia.

Mall, K.-G. (1984). Die Bedeutung der 129 Stoffe der EG-Liste für den Gewässerschutz, Z. Wasser-Abwasser-Forsch. **17,** 75–81.

Ministerium für Ernährung, Landwirtschaft, Umwelt und Forsten Baden-Württemberg, Wasserwirtschaftsverwaltung (1983). Leitfaden für die Beurteilung und Behandlung von Grundfwasserverunreinigungen durch leichtflüchtige Chlorkohlenwasserstoffe, Stuttgart.

Hellmann, H. (1982). Polycyclische aromatische Kohlenwasserstoffe in Acker- und Waldböden und ihr Beitrag zur Gewässerbelustung, Dtsch. Gewässerkd. Mitt. *26,* 63–69.

Hofius, K. (1981). Die sowjetische Weltwasserbilanz, Kurzbericht Dtsch. Gewässerkd. Mitt. **25,** 22–24.

VDI-Kommission Reinhaltung der Luft (1984). Schwermetalle in der Umwelt, Grundsatzstudie für das Bundesministerium des Innern, Düsseldorf.

OECD (1985). The State of the Environment 1985, Paris.

Index